Springer Series in Materials Science

Volume 238

The Springer Series in Materials Science covers the complete spectrum of materials physics, including fundamental principles, physical properties, materials theory and design. Recognizing the increasing importance of materials science in future device technologies, the book titles in this series reflect the state-of-the-art in understanding and controlling the structure and properties of all important classes of materials.

More information about this series at http://www.springer.com/series/856

Christopher R. Bowen · Vitaly Yu. Topolov
Hyunsun Alicia Kim

Modern Piezoelectric Energy-Harvesting Materials

Springer

Christopher R. Bowen
Department of Mechanical Engineering,
Materials Research Centre
University of Bath
Bath, Somerset
UK

Vitaly Yu. Topolov
Department of Physics
Southern Federal University
Rostov-on-Don
Russia

Hyunsun Alicia Kim
Department of Mechanical Engineering
University of Bath
Bath, Somerset
UK

and

Structural Engineering Department
University of California San Diego
San Diego, CA
USA

ISSN 0933-033X ISSN 2196-2812 (electronic)
Springer Series in Materials Science
ISBN 978-3-319-29141-3 ISBN 978-3-319-29143-7 (eBook)
DOI 10.1007/978-3-319-29143-7

Library of Congress Control Number: 2016931836

Printed on acid-free paper

This Springer imprint is published by SpringerNature
The registered company is Springer International Publishing AG Switzerland

To our colleagues, friends and pupils

Preface

> Energy is the capacity for doing work. It may exist in potential, kinetic, thermal, electrical, chemical, nuclear, or other various forms… Energy can be converted from one form to another in various ways. Usable mechanical or electrical energy is, for instance, produced by many kinds of devices, including fuel-burning heat engines, generators, batteries, fuel cells, and magnetohydrodynamic systems.
>
> Encyclopædia Britannica

> But do not try to keep what Providence
> Has given you, for your own use alone:
> We're doomed—and we all know it perfectly
> To squander not hoard the wealth we own.
>
> A. Akhmatova

An increasing interest in autonomous devices and electrical accumulators promoted by industrial and domestic applications has raised the issue of powering these systems. In the past decade, the important trend to address this problem consists in using ambient energy from the environment to supply autonomous devices and to make them self-powered with sufficient energetic concentration for various applications. Among the ambient-energy sources to be of interest, one can mention solar, thermal and mechanical sources. Despite this variety, much attention is paid to using mechanical energy, whose sources are available for small-size piezoelectric systems. Mechanical ambient energy to be taken from nature and converted into electrical energy is usually associated with strong winds, sea waves, sounds (acoustic waves) and earthquakes. A conversion of the mechanical form of energy into the electrical form implies using piezoelectric materials due to their sufficiently high energy densities, various electromechanical properties and performance in piezoelectric transducers, sensors, self-powered small-scale devices, hydrophones, etc.

In the past decade, rapid growth in the energy-harvesting field has been obvious, and thereafter the term '*piezoelectric energy harvesting*' has become widespread in the society of scientists and engineers. Piezoelectric energy harvesting is based on the direct piezoelectric effect at which electrical charge (or polarisation of a piezoelectric element) is generated from an external mechanical stress, strain, acoustic wave, vibration sources and so on. It should be added that the piezoelectric effect is concerned with one of three physical mechanisms of vibration-to-electric energy conversion. Along with the piezoelectric effect, electrostatic and electromagnetic transduction may be effective at this conversion, but to a lesser degree. A piezoelectric harvester can transform kinetic energy from mechanical vibrations into electrical energy due to the piezoelectric effect, and therefore, piezoelectric coefficients concerned with this effect should influence energy-harvesting characteristics of every piezoelectric harvester. Its important feature is that the geometric configuration and sizes of the piezoelectric harvester can be varied in wide ranges, and therefore, the system can be exploited on either a macro-scale or micro-scale level.

Among materials that meet conditions for piezoelectric energy-harvesting applications, in the first line we should highlight poled ferroelectric ceramics and composites based on either ferroelectric ceramics or relaxor-ferroelectric single crystals with high piezoelectric activity. Due to various adaptive characteristics, high piezoelectric performance and possibilities to vary and tailor their electromechanical properties in external fields, the aforementioned composites have been regarded as an important group of smart materials. Undoubtedly, complex and intricate interconnections between microstructure, composition and physical properties of the composites stimulate studies to predict and interpret the properties and related parameters of these materials under various conditions.

In the present monograph we discuss the piezoelectric performance and related parameters that are to be taken into consideration at piezoelectric energy harvesting. In particular, it provides a complex analysis of the *microgeometry–properties* relations in modern piezo-active composites, and this analysis broadens the traditional material-science concepts on the *composition–structure–property* relations and may be a stimulus to create novel high-effective materials with the predictable properties. Important examples of the piezoelectric performance of the composites or ceramics are discussed in the context of their anisotropy, piezoelectric sensitivity, electromechanical coupling, figures of merit and so on. The novelty of the monograph consists in the first systematisation of many authors' results on the performance of modern piezoelectric materials in the context of energy-harvesting applications.

The present monograph has been written on the basis of the authors' research results obtained at the Southern Federal University (Russia) and University of Bath (United Kingdom). The academic style of presentation of the research results and the discussion about these results indicate that this monograph would be useful to engineers, postgraduate students, researchers, and lecturers, i.e., to many specialists working in the field of smart materials, dealing with their effective electromechanical properties and applications. This monograph will be of benefit to all specialists looking to understand the anisotropic electromechanical properties and their links to piezoelectric energy harvesting, its parameters and applications. Some chapters and sections of the monograph may be a basis for a university course devoted to piezoelectric (or ferroelectric and related) materials and their energy-harvesting characteristics. Introducing this new monograph, we would like to mention lines from '*The Reader*' by A. Akhmatova as follows:

Each reader's a treasure-trove hidden

In fathoms of earth—even if

He is undistinguished, unbidden,

And has been mute all of his life.

And buried there, everything lies

That nature deems best to conceal…

Based on our knowledge, experience and new research results, we hope that the twenty-first century termed *The Century of New Materials and Technologies*, will lead to the fruitful development of new scientific directions in the field of piezoelectric energy harvesting and will promote creation of novel high-effective piezoelectric materials for energy-harvesting and related applications.

Bath, UK
Rostov-on-Don, Russia
San Diego, CA, USA

Christopher R. Bowen
Vitaly Yu. Topolov
Hyunsun Alicia Kim

Acknowledgements

The authors are grateful to Prof. Dr. A. Miles (University of Bath, UK), Prof. Dr. A. V. Turik, Prof. Dr. A.E. Panich, Prof. Dr. V.G. Gavrilyachenko, Prof. Dr. V. P. Sakhnenko, Prof. Dr. I.A. Parinov, Prof. Dr. A.A. Nesterov, Prof. Dr. A.A. Panich, Prof. Dr. V.L. Zemlyakov, Dr. V.V. Eremkin, Dr. V.G. Smotrakov, Dr. S. E. Filippov, Mr. A.A. Vorontsov, Mr. P.A. Borzov, Mr. I.A. Ermakov, and Ms. A.N. Isaeva (Southern Federal University, Russia), Prof. Dr. O. Kraft and Prof. Dr. M. Kamlah (Karlsruhe Research Centre, Germany), Prof. Dr. M. Lethiecq and Dr. F. Levassort (University of Tours, France), Dr. S.-H. Chang (National Kaohsiung Marine University, Taiwan, ROC), Prof. Dr. W. Cao (The Pennsylvania State University, USA), Prof. Dr. A. Safari and Dr. E.K. Akdogan (Rutgers—The State University of New Jersey, USA), Prof. Dr. A.S. Sidorkin (Voronezh State University, Russia), Prof. Dr. L.N. Korotkov and Prof. Dr. S.A. Gridnev (Voronezh State Technical University, Russia), and Dr. A.V. Krivoruchko (Don State Technical University, Russia) for their interest in the research problems and for their interesting and important research results that have been taken into account while writing the present book. The authors would like to thank Prof. Dr. P. Bisegna (University of Rome "Tor Vergata", Italy) for his activity in the field of finite element modelling and for his help at calculations of effective electromechanical properties of some piezo-active composites. The authors underline the vast geographic area wherein the piezoelectric and related energy-harvesting materials are studied or manufactured. The author Prof. Dr. C.R. Bowen would like to thank past supervisors including Prof. Dr. B. Derby (University of Manchester, UK) and Prof. Dr. N. Claussen (Technical University of Hamburg-Harburg, Germany). Professor Dr. C.R. Bowen would particularly like to acknowledge the kind support and advice of Prof. Dr. R. Stevens (University of Bath, UK); he will be sincerely missed.

The authors express much thanks to Prof. Dr. C. Ascheron and Mrs. E. Sauer (Springer-Verlag, Heidelberg, Germany) for their effective and timely cooperation in the field of editing and producing this book. Special thanks are extended to Prof. Dr. C. Ascheron for his invitation to write a new monograph on the performance of modern piezo-active materials. The authors sincerely thank Ms. R. Garner and Ms. C. Rothwell (University of Bath, UK) for their technical help. Copyright permissions obtained from Springer (www.springer.com), Elsevier (www.elsevier.com), IOP Publishing (www.iop.org), Taylor & Francis (www.informaworld.com), Wiley-VCH (www.wiley-vch.de), American Institute of Physics (www.aip.org), and The Royal Society of Chemistry (www.rsc.org) are acknowledged with due attention and gratitude.

Financial support that promoted the fruitful research collaboration and writing this book is acknowledged with many thanks. Hereupon, gratefully and proudly the authors mention the timely and effective support from the EPSRC (UK), National Physical Laboratory, QinetiQ (UK), Great Western Research (GWR, UK), University of Bath (UK), and Southern Federal University (Russia). Professor Dr. C.R. Bowen acknowledges funding from the European Research Council under the European Union's Seventh Framework Programme (FP/2007–2013)/ERC Grant Agreement no. 320963 on Novel Energy Materials, Engineering Science and Integrated Systems (NEMESIS). Dr. H.A. Kim acknolwedges support from the Engineering and Physical Science Research Council (EPSRC), Project EP/J014389/1.

Applied research has been performed with financial support from the Ministry of Education and Science of the Russian Federation (project RFMEFI57814X0088) using the equipment of the Centre of Collective Use 'High Technologies' at the Southern Federal University, and this support is acknowledged by the author, Prof. Dr. Vitaly Yu. Topolov, with gratitude.

Contents

Hyunsun Alicia Kim was born on 17 December 1974 in Seoul, Korea. She received a BEng (1997) and Ph.D. (2001) from the Department of Aeronautical Engineering, University of Sydney, Australia. Her Ph.D. research was in topology optimization, and its significance in the research field was recognised by the Zonta International Amelia Earhart Award. She continued the research in computational mechanics and structural optimisation at the University of Warwick (UK) and Virginia Tech (USA). She began her academic career as a Lecturer in 2001, subsequently Senior Lecturer then Reader in the Department of Mechanical Engineering, University of Bath, (UK). Since 2015, she is an Associate Professor in the Structural Engineering Department, University of California, San Diego (USA). She has been a visiting research scholar at NASA Langley and an affiliate of the Los Alamos National Laboratory.

Dr. H.A. Kim's primary research interests are topology optimization, computational mechanics and modelling, fibre composites, nonlinear porous materials, multifunctional structures and energy harvesting. She has over 150 peer-reviewed publications in international journals and conferences including best paper awards from the American Institute of Aeronautics and Astronics and the International Society of Structural and Multidisciplinary Optimization. Her research has been supported by Engineering and Physical Sciences Resarch Council (EPSRC), Royal Society, Leverhulme Trust, Airbus, and European Office of Aerospace Research & Development of the Air Force Office of Scientific Research and their support is gratefully acknowledged.

Chapter 1
The Piezoelectric Medium and Its Characteristics

An effect that links a mechanical action (mechanical stress or strain) with an electrical response (electric field, electric displacement or polarisation) is the piezoelectric effect or, more exactly, the *direct piezoelectric effect*. This effect was first studied by brothers P. Curie and J. Curie in experimental work (1880) on the behaviour of quartz single crystals (SCs) subjected to an external mechanical stress. The *converse piezoelectric effect* was later revealed in acentric dielectric SCs wherein an external electric field generated a mechanical response, i.e., a stress or strain of the sample, and this effect is similar to electrostriction. The piezoelectric effect follows a linear relationship between electric and mechanical variables and originates from the displacement of ions of an acentric SC under an applied electric field [1–4]. This relationship leads to a change in the sign of the piezoelectric effect when the direction of the external electric field is switched in the piezo-active medium. Such a medium is of interest due to the potential for a significant electromechanical coupling and conversion of energy from a mechanical form into electric one and vice versa.

In contrast to the *converse* piezoelectric effect, an electrostrictive strain (or stress) in any dielectric material follows a quadratic effect [3, 4] and does not switch sign under an alternating electric field. Electrostriction is caused by the electric field and cannot be represented as a direct or converse effect. It should be added that electrostriction is observed in any dielectric media, irrespective of the crystal symmetry and aggregate states (gas, liquid, SC, or amorphous).

Numerous experimental data show that, despite the linear character between electric and mechanical fields, the piezoelectric response of a SC sample is often intricate owing to various interconnections between the piezoelectric and other properties such as the elastic, dielectric (including pyro- and ferroelectric), magnetic and thermal properties [3]. The piezoelectric effect in poled ferroelectric ceramic (FC) and composite samples based on SCs or FCs is even more complex in comparison to the piezoelectric effect in ferroelectric SCs due to the influence of microstructure, domain orientations, intrinsic and extrinsic contributions, heterophase structures, etc. [5–11].

C.R. Bowen et al., *Modern Piezoelectric Energy-Harvesting Materials*,
Springer Series in Materials Science 238, DOI 10.1007/978-3-319-29143-7_1

In this chapter we consider some important characteristics of the piezoelectric medium in the context of its electromechanical coupling and energy-harvesting applications. We add that energy harvesting is currently of interest to both academia and industry since it provides a route for the development of autonomous and self-powered low-power electronic devices, e.g. for wireless sensor networks or consumable electronics.

1.1 Piezoelectric Coefficients

A general consideration of the piezoelectric effect in dielectric SCs is carried out in terms of thermodynamic functions, such as Helmholtz free energy, Gibbs free energy, elastic Gibbs energy, and electric Gibbs energy [2–4]. Each of these functions has at least three arguments that characterise the mechanical, electric and thermal states of the SC. The first argument can be either mechanical stress σ_{kl} or mechanical strain ξ_{jr}, the second argument can be either electric field $\boldsymbol{E}$ or electric displacement (electric flux density) $\boldsymbol{D}$, and the third argument can be either temperature T or entropy S. Obviously, these arguments are represented as second-rank (σ_{kl} and ξ_{jr}), first-rank (E_k and D_j) and zeroth-rank (scalars T and S) tensors. We may also develop additional arguments concerned, for instance, with an external magnetic field $\boldsymbol{H}$ or magnetic induction $\boldsymbol{B}$ [5].

Based on the thermodynamic functions and relations between the arguments of the mechanical, electric and thermal states [2, 3], one can describe the linked 'response–actions' in terms of ξ_{jr}, E_i and T as follows:

$$\sigma_{kl} = c^{E}_{kljr}\xi_{jr} - e_{ikl}E_i + \beta_{kl}\Delta T \quad (1.1)$$

$$D_i = e_{ikl}\xi_{kl} + \varepsilon^{\xi}_{if}E_f + p_i\Delta T \quad (1.2)$$

$$\Delta S = \beta_{jr}\xi_{jr} + p_iE_i + (\rho/T_0)c\Delta T \quad (1.3)$$

In (1.1)–(1.3) the constants are elastic moduli at constant electric field (c^{E}_{kljr}, fourth-rank tensor), piezoelectric coefficients (e_{ikl}, third-rank tensor), coefficients of thermal stress (β_{kl}, second-rank tensor), dielectric permittivities at constant mechanical strain (ε^{ξ}_{if}, second-rank tensor), pyroelectric coefficients (p_i, first-rank tensor, i.e., vector). In (1.1)–(1.3) density (ρ) and specific heat (c) of the SC are zeroth-rank tensors, i.e., scalars. Summing over the repeated subscripts [e.g., j, r and i in (1.1)] is to be performed from 1 to 3. The increments ΔT and ΔS in (1.1)–(1.3) denote differences $T - T_0$ and $S - S_0$, respectively, where T_0 is the initial temperature and S_0 is the initial entropy of the SC.

Linear relations from (1.1)–(1.3) hold in the case of relatively weak external fields [2–4]. For example, the linear dependence $\sigma_{kl}(\xi_{jr})$ from (1.1) obeys Hooke's law in an anisotropic medium [12] at small (<1 %) strains ξ_{jr}. The linear dependence $D_i(E_f)$ from (1.2) is valid at relatively low levels of electric field E_f applied to a dielectric SC. In the case of an acentric dielectric SC, a low level of electric field is one in which the E_f value is much smaller than the electric breakdown field. With regard to a ferroelectric SC that exhibit a piezoelectric performance due to the acentric crystal structure and spontaneous polarisation axis, similar requirements concerning the linear dependences $\sigma_{kl}(\xi_{jr})$ and $D_i(E_f)$ are possible, but the E_f range becomes narrower than that in the acentric linear dielectric SC. It is well known that in the presence of a low electric field E, the polarisation of a ferroelectric SC linearly depends on E [3, 4] so that domain-wall displacements are reversible and no nuclei of reoriented domains are formed. In this connection the electric field E_f is usually regarded to be a few times lower than the coercive field E_c which is determined from a ferroelectric hysteresis loop [3, 4].

Using (1.1)–(1.3), one can determine a set of isothermal constants (at T = const) of a piezoelectric SC. An analogous set of adiabatic constants (at S = const) can also be derived using thermodynamic functions and a combination of three arguments including ΔS (for example, ξ_{jr}, E_i and ΔS). According to experimental data, the difference between the related isothermic and adiabatic constants of the piezoelectric SCs is approximately 1 % or less [3, 5]. Based on this observation, the effect of thermal fields on the elastic and electric responses of the piezoelectric SCs is often neglected, and the description of the piezoelectric effect is carried out in terms of (1.1) and (1.2) at $\Delta T = 0$. Equations (1.1) and (1.2) are often represented in the following matrix form [2, 3]:

$$\sigma_p = c^E_{pq}\xi_q - e_{fp}E_f \tag{1.4}$$

$$D_k = e_{kl}\xi_l + \varepsilon^{\xi}_{kr}E_r \tag{1.5}$$

Equations (1.4) and (1.5) are written taking into account a conventional transition from two subscripts to one subscript in accordance with the well-known Vogt's rule [2–4, 12].

Equations (1.4) and (1.5) represent the first pair of piezoelectric equations that links two variables, mechanical strain ξ and electric field E. The first term in the right part of (1.5), $P_k = e_{kl}\xi_l$, describes the piezoelectric polarisation produced by an external mechanical strain as a result of the *direct* piezoelectric effect. In contrast to this, the *converse* piezoelectric effect is described by the $e_{fp}E_f$ term in (1.4), and the sequence of subscripts (fp, f) differs from the sequence (kl, l) at the direct piezoelectric effect.

From the thermodynamic treatment of the interrelations between the electric and elastic fields [2, 3], the converse and direct piezoelectric effects can also be described by the three following pairs of equations:

$$\xi_p = s^E_{pq}\sigma_q + d_{fp}E_f \tag{1.6}$$

$$D_k = d_{kl}\sigma_l + \varepsilon^{\sigma}_{kr}E_r \quad (1.7)$$

in variables of mechanical stress σ and electric field E,

$$\xi_p = s^D_{pq}\sigma_q + g_{fp}D_f \quad (1.8)$$

$$E_k = -g_{kl}\sigma_l + \beta^{\sigma}_{kr}D_r \quad (1.9)$$

in variables of mechanical stress σ and electric displacement D, and

$$\sigma_p = c^D_{pq}\xi_q - h_{fp}D_f \quad (1.10)$$

$$E_k = -h_{kl}\xi_l + \beta^{\xi}_{kr}D_r \quad (1.11)$$

in variables of mechanical strain ξ and electric displacement D. Superscripts σ and D denote measurement conditions at σ = const and D = const, respectively. Dielectric impermittivities β^{σ}_{kr} (at σ = const) and β^{ξ}_{kr} (at ξ = const) from (1.9) and (1.11) are determined [2, 3] using relations $\beta^{\sigma}_{kr}\varepsilon^{\sigma}_{rv} = \delta_{kv}$ and $\beta^{\xi}_{kr}\varepsilon^{\xi}_{rv} = \delta_{kv}$. Elastic moduli c^E_{pq} or c^D_{pq} and elastic compliances s^E_{pq} or s^D_{pq} are related by conditions $c^E_{pq}s^E_{qr} = \delta_{pr}$ and $c^D_{pq}s^D_{qr} = \delta_{pr}$, where δ_{kv} is the Kronecker symbol.

Equations (1.4)–(1.11) contain four types of the piezoelectric coefficients, namely e_{kl}, d_{kl}, g_{kl}, and h_{kl}, and in each pair of (1.4)–(1.11), there is a term corresponding to the direct piezoelectric effect and a term corresponding to the converse piezoelectric effect. Each piezoelectric coefficient characterises the relationship between components of two fields that are described by first- and second-rank tensors, and all the piezoelectric coefficients from (1.4)–(1.11) are components of the third-rank tensors, but written in the matrix form (i.e., with two subscripts) [2–5].

The piezoelectric coefficients e_{kl}, d_{kl}, g_{kl}, and h_{kl} can be used as an aid in the selection of piezoelectric materials for various piezotechnical applications. For example, for an actuator application [4], it may be necessary to select materials with a high strain per unit applied electric field (i.e., with large values of $|d_{kl}|$). For a pressure sensor, the electric field generated per unit mechanical stress is more likely to be important (i.e., with large values of $|g_{kl}|$).

The interrelationships of the piezoelectric coefficients d_{kl}, e_{kl}, g_{kl}, and h_{kl} follow from the thermodynamic description [2, 3] and are given by

$$d_{fp} = \varepsilon^{\sigma}_{fk}g_{kp} = e_{fq}s^E_{qp} \quad (1.12)$$

$$e_{fp} = \varepsilon^{\xi}_{fk}h_{kp} = d_{fq}c^E_{qp} \quad (1.13)$$

$$g_{fp} = \beta^{\sigma}_{fk}d_{kp} = h_{fq}s^D_{qp} \quad (1.14)$$

$$h_{fp} = \beta^{\xi}_{fk} e_{kp} = g_{fq} c^{D}_{qp} \quad (1.15)$$

The piezoelectric coefficients from (1.12)–(1.15) are also involved in relations [2, 3] between dielectric or elastic constants measured on different conditions:

$$\varepsilon^{\sigma}_{kr} - \varepsilon^{\xi}_{kr} = d_{kf} e_{rf} \quad (1.16)$$

$$\beta^{\xi}_{kr} - \beta^{\sigma}_{kr} = g_{kf} e_{rf} \quad (1.17)$$

$$c^{D}_{pq} - c^{E}_{pq} = e_{fp} h_{fq} \quad (1.18)$$

$$s^{D}_{pq} - s^{E}_{pq} = d_{fp} g_{fq} \quad (1.19)$$

As is seen from (1.12)–(1.19), there are close connections between the elastic, piezoelectric and dielectric constants of a piezoelectric medium. These connections enable us to term the sets of the elastic, piezoelectric and dielectric constants as electromechanical constants. These constants characterise electromechanical properties in a variety of materials, such as piezoelectric or ferroelectric SCs, poled FCs, thin ferroelectric films, piezo-active composites, etc.

As follows from (1.1), (1.2) and (1.4)–(1.11), the piezoelectric properties are described by third-rank tensors and are represented in the matrix form, as 3 × 6 or 6 × 3 matrices. Tables of the matrices of the piezoelectric coefficients related to acentric SCs, poled FCs, composites, and piezoelectric textures (i.e., to various piezoelectric media) have been collected in a series of monographs [2–5, 12]. As is known from crystallographic studies [1–3], the piezoelectric effect is found in 20 symmetry classes (or point groups) and 3 limiting symmetry classes (or Curie groups). The symmetry classes that obey the conditions for the piezoelectric effect [3, 12] are 1 (triclinic system), 2 and m (monoclinic system), 222 and $mm2$ (orthorhombic system), 3, 32 and $3m$ (trigonal or rhombohedral system), 4, 422, $4mm$, $\overline{4}$, and $\overline{4}2m$ (tetragonal system), 6, 622, $\overline{6}$, and $\overline{6}m2$ (hexagonal system), and 23 and $\overline{4}3m$ (cubic system). The Curie groups that satisfy conditions for the piezoelectric texture [3, 12] are ∞, ∞mm and $\infty/2$. For instance, in the 1 class, the $||d||$ matrix comprises the largest number of independent constants, i.e., d_{11}, d_{12}, …, d_{16}, d_{21}, d_{22}, …, d_{26}, d_{31}, d_{32}, …, d_{36}, or 18 piezoelectric coefficients. These piezoelectric coefficients are written in the form as shown in (1.12) and (1.13).

The conditions for measuring the piezoelectric coefficients are associated with the electric and mechanical variables from (1.4)–(1.11). If the direct piezoelectric effect is considered, then the piezoelectric coefficients of the medium are measured [3] using the following relations for the electric polarisation P_k and field E_k:

(i) $P_k = d_{kl}\,\sigma_l$ (for d_{kl}),
(ii) $E_k = -g_{kl}\,\sigma_l$ (for g_{kl}),
(iii) $E_k = -h_{kl}\,\xi_l$ (for h_{kl}), and
(iv) $P_k = e_{kl}\,\xi_l$ and $E = 0$ (for e_{kl}).

We remind that $P_k = D_k$ at $E_k = 0$, and the polarisation P_k is determined from a surface density of electric charge on a sample face [3]. At the converse piezoelectric effect, the following relations are taken into account:

(i) $\xi_p = d_{fp} E_f$ (for d_{fp}),
(ii) $\xi_p = g_{fp} D_f$ (for g_{fp}),
(iii) $\sigma_p = -h_{fp} D_f$ (for h_{fp}), and
(iv) $\sigma_p = -e_{fp} E_f$ (for e_{fp}).

As follows from the above-given conditions, it is relatively easy to find e_{kl} and h_{kl} as constants of the direct piezoelectric effect and d_{kl} as a constant of the converse piezoelectric effect [3].

1.2 Electromechanical Coupling at Various Oscillation Modes

1.2.1 Energy Conversion

Full sets of elastic, piezoelectric and dielectric constants from (1.4)–(1.11) enable us to estimate the effectiveness of the conversion of electric energy into mechanical energy and vice versa. As follows from the analysis of the energy conversion [2, 13] in a piezoelectric medium, this effectiveness depends on differences between dielectric permittivities from (1.16) or differences between elastic compliances from (1.19), and these differences strongly depend on the piezoelectric effect. In a general form, the effectiveness of the energy conversion is described by an electromechanical coupling factor (ECF) [2, 3, 5, 6, 13]

$$k = w_{piezo} / \sqrt{w_{el} w_{mech}} \tag{1.20}$$

and is expressed in terms of volume densities of piezoelectric (or mutual) energy $w_{piezo} = \sigma_i \ d_{mi} \ E_m/2$, electric energy $w_{el} = E_l \ \varepsilon_{lr}^{\sigma} \ E_r/2$, and mechanical energy $w_{mech} = \sigma_p \ s_{pq}^{E} \sigma_q/2$. The k^2 value calculated from (1.20) characterises [2–5, 9, 13] a ratio of stored mechanical energy to electric energy input (when a portion of electric energy is applied to the piezoelectric sample) or a ratio of stored electric energy to mechanical energy input (when a portion of mechanical energy is applied to the piezoelectric sample). The k^2 value also characterises a measure of the magnitude of the piezoelectric transducer bandwidth [2]. The absolute ECF value depends [2, 4, 9, 13] on the vibration mode, measurement methodology, sample shape, and the electromechanical constants of the piezoelectric medium. As a rule, two kinds of ECFs are considered as follows: the static ECF determined from equations of the state of the piezoelectric medium and the dynamic (effective) ECF related to the individual piezoelectric element [9, 13]. It is obvious that the ECFs strongly depend on the orientation of the external electric or mechanical field applied to the piezoelectric element and on the symmetry of its properties.

1.2.2 Examples of Electromechanical Coupling Factors

A poled FC element with a remanent polarisation vector $\boldsymbol{P_r} \parallel OX_3$ (∞mm symmetry) is characterised [2–5] by three independent piezoelectric coefficients, namely, d_{31}, d_{33} and d_{15}. These piezoelectric coefficients are linked to the ECFs as follows:

$$k_{33} = d_{33}/\left(\varepsilon_{33}^{\sigma} s_{33}^{E}\right)^{1/2} \tag{1.21}$$

(ECF at the longitudinal oscillation mode, or simply, longitudinal ECF),

$$k_{31} = d_{31}/\left(\varepsilon_{33}^{\sigma} s_{11}^{E}\right)^{1/2} \tag{1.22}$$

(ECF at the transverse oscillation mode) and

$$k_{15} = d_{15}/\left(\varepsilon_{11}^{\sigma} s_{55}^{E}\right)^{1/2} \tag{1.23}$$

(ECF at the shear oscillation mode). Since an equality $s_{55}^E = s_{44}^E$ holds [2–5] in a poled FC medium, expression (1.23) is often given by $k_{15} = d_{15}/(\varepsilon_{11}^{\sigma} s_{44}^{E})^{1/2}$. For various piezo-active materials the ECF is represented in the general form as $k_{ij} = d_{ij}/(\varepsilon_{ii}^{\sigma} s_{ij}^{E})^{1/2}$, and hereby it is sufficient to require $d_{ij} \neq 0$ at a specific oscillation mode. Due to equalities $s_{22}^E = s_{11}^E$ and $d_{24} = d_{15}$, which are valid for any FC poled along the OX_3 axis [2–5], the relations

$$k_{32} = k_{31} \text{ and } k_{24} = k_{15} \tag{1.24}$$

are also valid. It should be noted that a violation of conditions (1.24) may be accounted for by peculiarities of symmetry of the piezoelectric medium. For example, in a piezoelectric material belonging to the $mm2$ class, we should distinguish [3] s_{22}^E and s_{11}^E, d_{24} and d_{15}, etc. As a consequence, inequalities $k_{32} \neq k_{31}$ and $k_{24} \neq d_{15}$ hold.

Along with the ECFs from (1.21)–(1.24), the following ECFs are considered widely [2, 4, 9, 10, 13] and often measured on poled FC and piezo-active composite samples with ∞mm symmetry:

$$k_t = e_{33}/\left(c_{33}^{D} \varepsilon_{33}^{\xi}\right)^{1/2} \tag{1.25}$$

(ECF at the thickness oscillation mode, or simply, thickness ECF),

$$k_p = k_{31}[2/(1-\sigma_E)]^{1/2} \tag{1.26}$$

(ECF at the planar oscillation mode, or simply, planar ECF) and

$$k_h = d_h / \left(\varepsilon_{33}^{\sigma} s_h^E\right)^{1/2} \tag{1.27}$$

(hydrostatic ECF). In (1.26) σ_E is the Poisson's ratio at electric field E = const, and in (1.27) d_h and s_h^E are the hydrostatic piezoelectric coefficient and hydrostatic elastic compliance at E = const, respectively. Tables of ECFs measured on the FC elements at specific oscillation modes are given in work [2, 4, 13]. Data on the ECFs from (1.21)–(1.27) along with the piezoelectric coefficients from (1.12)–(1.15) are useful in the selection of piezoelectric materials for active elements of electromechanical transducers, hydrophones and piezoelectric energy harvesters [10, 13–24].

It should be added that examples of ECFs and their anisotropy in piezoelectric materials will be discussed in Chap. 2.

1.3 Figures of Merit

According to Uchino and Ishii [19], there are three major stages that are concerned with piezoelectric energy harvesting in electromechanical systems. These stages are characterised as follows:

(i) mechanical-mechanical energy transfer, including the mechanical stability of the piezoelectric transducer under a large mechanical stress, and mechanical impedance matching,
(ii) mechanical-electric energy transduction due to electromechanical coupling in a piezoelectric element (SC, ceramic or composite) and
(iii) electric-electric energy transfer, including electric impedance matching, for instance, in the presence of a DC/DC converter to accumulate the energy into an electric rechargeable battery.

Mechanical (acoustic) impedance matching is one of the important factors to be taken into account at the first stage [19]. The mechanical impedance of a material is given by $Z = (\rho c_{ij})^{1/2}$, where ρ is its density, and c_{ij} is its elastic modulus. At the second stage, a portion of mechanical energy that reaches the piezoelectric transducer is converted into electric energy due to electromechanical coupling. Hereby a voltage induced in this transducer is written in the general form as

$$V = g_{ij} F t / A \tag{1.28}$$

In (1.28) g_{ij} is the piezoelectric coefficient, F is the applied force, t is the thickness of the piezoelectric element, and A is the area of the surface on which the force acts. As a result of mechanical loading and due to the piezoelectric effect, an output electric power is

$$P = (CV^2/2)f \tag{1.29}$$

In (1.29) C is the electric capacitance of the piezoelectric element, and f is frequency of its vibration. We note that the capacitance C is proportional to dielectric permittivity $\varepsilon_{jj}^{\sigma}$ for a flat large electric condensator, and f is chosen in a low-frequency range far from a resonance frequency of this piezoelectric element.

When the mechanical loading is along the poling axis OX_3 (as is conventional for FCs [2–5]) with electrodes that are perpendicular to OX_3, the output electric power is $P_{out} \sim d_{33}g_{33}$, and

$$(Q_{33})^2 = d_{33}g_{33} \tag{1.30}$$

is the squared figure of merit [10, 14, 19, 24] concerned with the oscillation '33' (or longitudinal) mode. Taking into account (1.28) and (1.30), one can represent energy generated in the piezoelectric transducer at the applied force F as $W_{gen} = (Q_{33})^2 F^2 t/(2lw)$, where l is the length, and w is the width of the transducer in the form of the rectangular parallelepiped.

Along with $(Q_{33})^2$ related to the longitudinal piezoelectric effect, we introduce squared figures of merit

$$(Q_{31})^2 = d_{31}g_{31} \text{ and } (Q_{32})^2 = d_{32}g_{32} \tag{1.31}$$

which are related to the transverse piezoelectric effect and oscillation '31' and '32' modes, respectively, and

$$(Q_h)^2 = d_h g_h \tag{1.32}$$

that is a hydrostatic analog of the squared figure of merit from (1.30).

Squared figures of merit (1.30)–(1.32) are often used to characterise the sensor signal-to-noise ratio of the piezoelectric material and its sensitivity [10, 25]. A piezoelectric element with a large value of $(Q_{3j})^2$ from (1.30) or (1.31) will generate high voltage and power when this element is exploited for energy harvesting and sensing applications. It should be added that the electrical damping plays an important role in affecting the performance of the piezoelectric element. Taking into account dielectric losses (tan δ) in the low-frequency region, the squared figure of merit in an off-resonance region may be represented [24] as $(Q_{31})^2 = d_{31}g_{31}/\tan\delta$ (for a 3–1 mode of the transducer), $(Q_{32})^2 = d_{32}g_{32}/\tan\delta$ (for a 3–2 mode of the transducer) or $(Q_{33})^2 = d_{33}g_{33}/\tan\delta$ (for a 3–3 mode of the transducer). The criteria for maximising the $(Q_{33})^2$ in poled FCs and trends in optimising the FC composition for energy-harvesting related applications were considered in works [20, 23].

Numerous examples of figures of merit and related parameters of piezoelectric materials will be discussed in Chap. 3.

1.4 Hydrostatic Parameters Concerned with Electromechanical Coupling

In Sects. 1.2.2 and 1.3 we introduced some of the hydrostatic parameters of a piezoelectric medium, for instance, the hydrostatic piezoelectric coefficients d_h and g_h, electromechanical coupling factor k_h and squared figure of merit $(Q_h)^2$. Taking into account the symmetry features of a piezoelectric element and specific orientations of its electrode structure, we may consider different relationships for d_h:

(i) $d_h = d_{33} + d_{32} + d_{31}$ at electrodes perpendicular to OX_3,
(ii) $d_h = d_{23} + d_{22} + d_{21}$ at electrodes perpendicular to OX_2 and
(iii) $d_h = d_{13} + d_{12} + d_{11}$ at electrodes perpendicular to OX_1.

Similar relationships are valid for the hydrostatic piezoelectric coefficients d_h and e_h. A complicated system of electrodes can lead to an increase in d_h due to positive contributions from different co-ordinate directions OX_j. An additional route for increasing d_h, g_h and e_h is the use of piezo-active composites [26] wherein microgeometric factors and polarisation orientation effects can promote favourable relations between the piezoelectric coefficients related to the response along the OX_j axes. Finally, the hydrostatic elastic compliance s_h^E from (1.27) characterises an elastic response of the piezoelectric element under a hydrostatic pressure. In general form s_h^E is represented as $s_h^E = s_{11}^E + s_{22}^E + s_{33}^E + 2(s_{12}^E + s_{13}^E + s_{23}^E)$, where s_{ab}^E is the elastic compliance at E = const.

1.5 Effective Electromechanical Properties in Heterogeneous Piezoelectric Media

1.5.1 Relaxor-Ferroelectric Domain-Engineered Single Crystals

Relaxor-ferroelectric perovskite-type solid solutions are materials with one of the following general formulae: $(1-x)Pb(B_1, B_2)O_3 - xPbTiO_3$ or $(1-x-y)Pb(B_1, B_2)O_3 - xPb(B_1', B_2')O_3 - yPbTiO_3$. The complex perovskites described with the general formula $Pb(B_1, B_2)O_3$ [or $Pb(B_1', B_2')O_3$] are disordered dielectric compounds that contain ions of metals from the following groups: B_1 (or B_1') = Mg, Zn, Ni, Fe, Sc, Yb, and In (low-valence metals) and B_2 (or B_2') = Nb, Ta and W (high-valence metals) [27, 28]. A combination of metal ions with low and high valences results in physical properties that distinguish $Pb(B_1, B_2)O_3$ from 'normal' (ordered or regular) perovskite-type ferroelectrics such as $PbTiO_3$, $BaTiO_3$ or $KNbO_3$. The complex perovskites $Pb(B_1, B_2)O_3$ exhibit a broad and frequency-dispersive dielectric maxima and contain polar nanoregions (with ferroelectric or antiferroelectric ordering) in a non-polar phase over a wide temperature range. These materials are characterised by a relaxation dielectric polarisation and are often termed *relaxors* or *ferroelectric relaxors* [27–29].

The perovskite-type solid solutions of $(1 - x)Pb(B_1, B_2)O_3 - xPbTiO_3$ combine the physical properties of the relaxor-type and 'normal' ferroelectric components, and, as a rule, excellent electromechanical properties are observed near the morphotropic phase boundary [27, 30, 31]. To achieve a high piezoelectric activity in SC samples, the relaxor-ferroelectric solid solutions are often engineered by compositional adjustment with a corresponding decrease in Curie temperature of the paraelectric-to-ferroelectric phase transition [27, 31, 32], and specific domain-engineered structures are formed in an electric field [31, 32].

Since the 2000s, domain-engineered SCs of $(1 - x)Pb(Mg_{1/3}Nb_{2/3})O_3 - xPbTiO_3$ (PMN–*x*PT) and $(1 - x)Pb(Zn_{1/3}Nb_{2/3})O_3 - yPbTiO_3$ (PZN–*y*PT) have been in the focus of many experimental and theoretical studies. Of particular interest are compositions taken in the vicinity of the morphotropic phase boundary, with intricate domain/heterophase states [33] etc. Results of numerous studies show that the engineered non-180° domain structures [31, 32, 34], intermediate ferroelectric phases [30, 33, 35] and domain-orientation processes [36–39] play an important role in forming outstanding electromechanical properties in the aforementioned relaxor-ferroelectric SCs. Due to these and other phenomena, the domain-engineered PMN–*x*PT and PZN–*y*PT SCs poled along the certain crystallographic directions (often along [001], [011] or [111] of the perovskite unit cell) [27, 31] exhibit very high piezoelectric activity and significant electromechanical coupling [28, 31, 34, 36], which is of benefit for energy-harvesting applications.

Full sets of electromechanical constants of domain-engineered relaxor-ferroelectric SCs with compositions near the morphotropic phase boundary are given in Table 1.1. We note that the electromechanical properties are regarded as effective, i.e., averaged over a macroscopic volume of a SC sample with fixed orientations of domains and domain walls separating them. Typical absolute values

Table 1.1 Experimental values of elastic compliances s^E_{ab} (in 10^{-12} Pa^{-1}), piezoelectric coefficients d_{ij} (in pC/N) and dielectric permittivities ε^σ_{pp} of [001]-poled domain-engineered PMN–*x*PT and PZN–*y*PT SCs (4*mm* symmetry) at room temperature

SC	s^E_{11}	s^E_{12}	s^E_{13}	s^E_{33}	s^E_{44}	s^E_{66}	d_{31}	d_{33}	d_{15}	$\varepsilon^\sigma_{11}/\varepsilon_0$	$\varepsilon^\sigma_{33}/\varepsilon_0$
PMN–0.33PT [59]	69.0	−11.1	−55.7	119.6	14.5	15.2	−1330	2820	146	1600	8200
PMN–0.30PT [60]	52.0	−18.9	−31.1	67.7	14.0	15.2	−921	1981	190	3600	7800
PMN–0.28PT [61]	44.57	−28.91	−13.91	34.38	15.22	16.34	−569	1182	122	1672	5479
PZN–0.045PT [62]	82.0	−28.5	−51.0	108	15.6	15.9	−970	2000	140	3100	5200
PZN–0.07PT [63]	85.9	−14.1	−69.0	142	15.9	14.1	−1204	2455	176	3000	5622
PZN–0.08PT [63]	87.0	−13.1	−70.0	141	15.8	15.4	−1455	2890	158	2900	7700

of piezoelectric coefficients $|d_{3j}| \sim 10^3$ pC/N in the domain-engineered PMN–*x*PT and PZN–*y*PT SCs (Table 1.1) are larger than the $|d_{3j}|$ of regular ferroelectric single-domain $PbTiO_3$ SC [40] and conventional FC based on $Pb(Zr_{1-x}Ti_x)O_3$ [4, 5, 9, 10, 13]. Moreover, the piezoelectric coefficient d_{33} in the [001]-poled heterophase PZN–0.08PT SC can be approximately 12,000 pC/N [35] due to the electric-field-induced phase transition and the presence of an intermediate monoclinic phase.

Experimental results [28, 32] suggest that relatively high piezoelectric strains (over 0.5 %) are achieved in the [001]-poled domain-engineered PZN–*y*PT SCs under an electric field $\boldsymbol{E}$ || [001]. At room temperature these SCs are characterised by 3*m* symmetry in the single-domain state and by 4*mm* symmetry in the poly-domain state wherein the spontaneous polarisation vectors of the non-180° domains are parallel to the following perovskite unit-cell directions: [111], [$\bar{1}$ 11], [1 $\bar{1}$ 1], and [$\bar{1}$ $\bar{1}$ 1] [27, 31, 41].

Important interconnections between the structure and properties in the PMN–*x*PT and PZN–*y*PT SCs have been reviewed by Noheda [30]. The high piezoelectric activity in these SCs is associated with a polarisation rotation in the electric field $\boldsymbol{E}$ [36, 38]. This rotation between the single-domain states in the tetragonal (4*mm*) and rhombohedral (3*m*) phases can be implemented in different ways that form the intermediate monoclinic phases [30] and complex heterophase states [33, 35] near the morphotropic phase boundary.

The relaxor-ferroelectric SCs are used as active elements of energy harvesters [42–45] and as piezoelectric components of advanced composites [46] that are suitable for energy-harvesting devices due to the large piezoelectric coefficients d_{3j}, ECF k_t and other parameters.

1.5.2 Poled Ferroelectric Ceramics

Full sets of the electromechanical constants of ferroelectric polydomain SCs are used to study the relationships between electromechanical properties in a single grain and a poled FC [9] consisting of a large number of grains with different orientations of main crystallographic axes. This study is based on solving a problem [7, 9, 47–49] of the electromechanical interaction between a spherical piezo-active grain and a piezo-ceramic medium surrounding this grain. Such an interaction is considered within the framework of the effective medium method [48, 49] that takes into account not only the piezoelectric effect but also the anisotropy of the piezoelectric properties within the grain and the FC medium. This anisotropy can be considerable, particularly in the presence of non-180° domain reorientations in the grains [50]. The effect of the 90° domain-wall displacements within the grains on the electromechanical properties in the poled $BaTiO_3$ FC has been studied [7, 51, 52], and contributions from these domain-wall displacements to a variety of constants for the FC medium were evaluated.

In contrast to the $BaTiO_3$ FC, the electromechanical properties in the $PbTiO_3$ FC are determined [42, 53] in the absence of 90° domain-wall displacements within grains. According to results [53, 54], the 90° domain structure of the grains has a strong influence on the piezoelectric coefficients d_{fp} and their anisotropy d_{33}/d_{31} in $PbTiO_3$-type FCs. In particular, it is possible to attain a large piezoelectric anisotropy (i.e., $d_{33}/|d_{31}| \gg 1$) by doping the FC, varying the temperature and by changing the volume fractions of the 90° domains within ceramic grains [54].

Subsequent averaging procedures facilitate consideration of the hierarchy-of-properties chain of 'single-domain SC—polydomain SC—poled FC' in the presence of the 90° domain-wall displacements ($BaTiO_3$) or assuming that the 90° domain walls are motionless ($PbTiO_3$) [42, 53, 54]. Knowledge of the full sets of the electromechanical constants of ferroelectric ceramics enables appropriate material selection for various piezotechnical applications, such as energy harvesting, and offers a route to using poled FCs as piezo-active components for composite materials.

The electromechanical constants of poled FC materials are often measured using standard methods [55, 56] at weak electric fields. The majority of the FCs are manufactured as perovskite-type solid solutions [3–5, 9, 13, 57, 58] with compositions varied and tailored to improve specific properties. Attempts to classify the FCs have been made in a variety of papers [9, 57, 58]. From the classification scheme [57] and by considering the conventional perovskite-type ceramics based on $Pb(Zr_{1-x}Ti_x)O_3$, it is possible to identify materials which are stable with regard to electric and mechanical loads, materials with high dielectric permittivity $\varepsilon_{33}^{\sigma}$, materials having high sensitivity with respect to mechanical stress fields (large absolute values of piezoelectric coefficients g_{fp}), materials with the large piezoelectric anisotropy d_{33}/d_{31}, materials with high stability with regard to resonance frequency, materials with low dielectric permittivity $\varepsilon_{33}^{\sigma}$, and high-temperature materials (high Curie temperature T_C). From the classification method proposed in monograph [9], the $Pb(Zr_{1-x}Ti_x)O_3$-based FCs are divided into four groups with specific performances. These groups are related to molar concentrations of Ti from molar-concentration ranges $0 \leq x \leq 0.1$, $0.1 \leq x \leq 0.4$, $0.4 \leq x \leq 0.6$, and $0.6 \leq x \leq 1$, and some advantages of each group of poled FC materials have been discussed [9] in the context of piezotechnical applications.

Examples of the full sets of electromechanical constants of poled FC materials are given in Table 1.2. It is seen that the piezoelectric coefficients d_{3j} of the listed FCs are considerably smaller than the piezoelectric coefficients d_{3j} of well-known domain-engineered relaxor-ferroelectric SCs (Table 1.1), i.e., piezoelectric activity of the FC medium along the poling axis is less pronounced. This feature can be associated with the grain microstructure of the FC, with internal mechanical stress fields in each grain and with restricted possibilities of domain reorientations therein.

We now compare some parameters of relaxor-ferroelectric SCs to the related parameters of poled FCs. Among the parameters of interest for energy harvesters, we mention ECFs k_{3j} and squared figures of merit $(Q_{3j})^2$. Published data from Table 1.3 suggests that the electromechanical properties of the PMN–xPT and

Table 1.2 Experimental values of elastic compliances s_{pq}^E (in 10^{-12} Pa^{-1})[a], piezoelectric coefficients d_{ij} (in pC/N) and dielectric permittivities $\varepsilon_{kk}^{\sigma}$ of perovskite-type FCs at room temperature

FC	s_{11}^E	s_{12}^E	s_{13}^E	s_{33}^E	s_{44}^E	d_{31}	d_{33}	d_{15}	$\frac{\varepsilon_{11}^{\sigma}}{\varepsilon_0}$	$\frac{\varepsilon_{33}^{\sigma}}{\varepsilon_0}$
$BaTiO_3$ (I) [70]	8.55	−2.61	−2.85	8.93	23.3	−79	191	270	1623	1900
$BaTiO_3$ (II) [13]	9.1	−2.7	−2.9	9.5	22.8	−78	190	260	1450	1700
ZTS-19 [5]	15.1	−5.76	−5.41	17.0	41.7	−126	307	442	1350	1500
PZT-4 [13]	12.3	−4.03	−5.35	15.6	39.1	−124	291	496	1440	1280
PZT-5 [13]	16.3	−5.67	−7.17	18.7	47.4	−170	373	583	1730	1700
PZT-5H [71]	10.8	−3.28	−3.41	11.6	28.3	−128	315	482	2640	2490
PZT-7A [72]	10.7	−3.22	−4.62	13.9	39.5	−60.2	151	364	843	427
PZ 27 [73, 74]	16.9	−6.32	−8.56	22.5	43.9	−174	419	515	1800	1770
PZ 34 [73, 74]	7.71	−1.54	−3.99	13.0	17.0	−3.92	59.1	39.2	191	196
PCR-1, hp[b] [57]	12.5	−4.4	−5.8	15.9	38.8	−95	220	420	1130	650
PCR-7, hp [57]	17.2	−7.2	−6.7	17.3	42.4	−280	610	760	2970	3500
PCR-7M, hp [57]	17.5	−6.7	−7.9	19.6	43.8	−350	760	880	3990	5000
PCR-8, hp [57]	12.5	−4.6	−5.2	15.6	35.3	−130	290	410	1380	1400
PCR-8, ct [57]	13.3	−4.8	−4.4	14.5	39.5	−125	280	458	1320	1300
PCR-13, ct [57]	10.4	−3.7	−2.1	11.3	28.5	−65	140	200	870	780
PCR-21, hp [57]	11.8	−4.5	−3.9	12.6	40.8	−109	250	370	1400	1350
PCR-63, hp [57]	9.8	−3.5	−2.7	9.8	24.1	−60	140	166	960	1170
PCR-73, hp [57]	17.9	−6.8	−9.6	23.5	43.7	−380	860	980	4750	6000
$Pb(Zr_{0.54}Ti_{0.46})O_3$ [13]	11.6	−3.33	−4.97	14.8	45.0	−60.2	152	440	990	450
$Pb(Zr_{0.52}Ti_{0.48})O_3$ [75]	13.8	−4.07	−5.80	17.1	48.2	−93.5	223	494	1180	730
$(Pb_{0.94}Sr_{0.06})(Ti_{0.47}Zr_{0.53})O_3$ [75]	12.3	−4.05	−5.31	15.5	39.0	−123	289	496	1475	1300
Modified $PbTiO_3$ (I) [76]	7.50	−1.51	−1.10	8.00	17.9	−4.40	51.0	53.0	228	177
Modified $PbTiO_3$ (II) [77]	7.7	−1.7	−1.2	8.2	19	−6.8	56	68	240	190
$(Pb_{0.9625}La_{0.025})\cdot$ $\cdot(Ti_{0.99}Mn_{0.01})O_3$ [78]	7.20	−1.42	−1.73	7.62	15.8	−4.15	47.2	53.0	223	168
$(Pb_{0.85}Nd_{0.10})(Ti_{0.99}Mn_{0.01})\cdot$ $\cdot O_3$ [78]	6.84	−1.50	−1.83	7.27	16.4	−5.42	56.8	79.5	313	252
$(Pb_{0,855}Nd_{0,11})\cdot$ $\cdot(Ti_{0,94}Mn_{0,02}In_{0,04})O_3$ [78]	6.77	−1.55	−1.79	7.29	16.0	−6.40	57.9	86.4	318	248
PMN– 0.35PT [79]	13.2	−3.96	−6.05	14.7	33.4	−133	270	936	4610	3270

[a]In the poled state of any FC (∞mm symmetry, the poling axis is parallel to the co-ordinate axis OX_3), elastic compliance s_{66} is determined [2, 3, 9, 12, 13] as follows: $s_{66} = 2(s_{11}^E - s_{12}^e)$

[b]FC samples of the PCR type have been manufactured using either the conventional technology (ct) or hot pressing (hp). PCR is the abbreviation for the group 'piezoelectric ceramics from Rostov-on-Don' (Russia) [57]

PZN–yPT SCs lead to larger values of k_{33}, $|k_{31}|$ and $(Q_{3j})^2$ in comparison to those of various FCs. This is a result of the high piezoelectric activity of the relaxor-ferroelectric SCs and the key role of the piezoelectric coefficients d_{3j} in the formation of k_{3j} and $(Q_{3j})^2$ [see (1.21), (1.22), (1.30), and (1.31)]. It should be added for a further comparison that textured PMN–$PbZrO_3$–$PbTiO_3$ FC samples studied in work [22] are characterised by $(Q_{33})^2 = 59.18 \cdot 10^{-12}$ Pa^{-1}, and this value

Table 1.3 ECFs k_{3j} and squared figures of merit $(Q_{3j})^2$ (in 10^{-12} Pa^{-1}) of some relaxor-ferroelectric SCs and poled FCs (evaluated using data from Tables 1.1 and 1.2)

Material	k_{31}	k_{33}	$(Q_{31})^2$	$(Q_{33})^2$
Domain-engineered SCs				
PMN–0.33PT	–0.594	0.957	24.4	110
PMN–0.30PT	–0.486	0.916	12.3	56.9
PMN–0.28PT	–0.387	0.915	6.66	28.8
PZN–0.045PT	–0.499	0.897	20.4	86.9
PZN–0.07PT	–0.582	0.924	29.1	121
PZN–0.08PT	–0.598	0.932	31.1	123
Poled FCs				
PZT-5	–0.343	0.703	1.92	9.25
PZT-5H	–0.262	0.623	0.743	4.50
PZT-7A	–0.299	0.659	0.959	6.03
PCR-1	–0.126	0.727	1.57	8.41
PCR-7	–0.384	0.833	2.53	12.0
PCR-7M	–0.398	0.816	2.77	13.1
Modified $PbTiO_3$ (I)	–0.0406	0.456	0.0124	1.66

is comparable to that related to the domain-engineered PMN–0.30PT SC and much more than $(Q_{33})^2$ of conventional FCs (see Table 1.3). We add that examples of the squared figures of merit $(Q_{3j})^2$ and $(Q_h)^2$ of poled perovskite-type FCs are also discussed in Sect. 3.1.

The piezoelectric and elastic anisotropy of FCs leads to larger ratios of $k_{33}/|k_{31}|$ and $(Q_{33})^2/(Q_{31})^2$ which are of particular importance for energy-harvesting applications. The relevant piezoelectric element will enable us to transform energy on the specific direction, and the main energy flow, its transformation and piezoelectric sensitivity of this element will be along the poling axis OX_3.

1.5.3 *Piezo-Active Composites*

Composites are heterogeneous systems that consist of two or more components that differ in chemical composition and properties and are separated by distinct interfaces [5]. Each composite material is characterised by the following features [5, 10]:

(i) the typical size of the separate structural elements are small in comparison with the whole composite sample (i.e., there is an element of micro-inhomogeneity) and
(ii) the typical sizes of the structural elements are greater than the size of their individual atoms or molecules so that each component of the composite sample is regarded as a continuous medium. As a result, to describe the composite properties, it is possible to consider the physical laws and equations suitable for continuous media.

The piezo-active composites (often termed *piezo-composites*) form an important group of modern smart materials. This group is vast due to the large number of components that may be involved in the design of the composites. Among the wide range of piezo-active components to be considered, ferroelectrics (FCs and SCs) play an important role [4, 5, 10, 26]. Moreover, modern processing technologies enable the synthesis of a variety of ferroelectric SCs, FCs, ferroelectric films, and nano-sized ferroelectrics. The ferroelectric properties of these systems are generally related to the change of the spontaneous polarisation vector in an external electric field (domain switching).

Piezo-active composites are of great interest due to the ability to vary and tailor the microgeometry, effective physical properties, their anisotropy, and hydrostatic and other parameters across a wide range [5, 10, 26]. These significant variations appear upon combining different components [5, 10], for example,

(i) FC and polymer (the latter can be either piezo-passive or piezo-active),
(ii) ferroelectric SC and FC, or
(iii) ferroelectric SC and polymer.

In various compositions (i)–(iii) at least one of the components always exhibits some piezoelectric properties. The presence of one or more piezoelectric components enables a set of parameters or properties to be attained that are not specific to the separate components, and this synergy is important for piezoelectric transducer, hydroacoustic and other applications [10, 26]. The piezo-active composites based on ferroelectrics form the final (and very important) link in hierarchy-of-properties chains [10, 26, 54] of 'single-domain SC → polydomain SC → FC → composite' and 'single-domain SC → polydomain SC → heterophase SC → composite'. We add that since the 1980s, FCs have been of interest as piezoelectric components of composites [4, 5, 10]. However since the 2000s, relaxor-ferroelectric SCs with high piezoelectric activity [59–63] are also used as components of modern piezo-composites (see, for instance, [19, 26, 46]). Due to their remarkable electromechanical properties, the piezo-active composites form an important group of modern smart materials [10], i.e., materials that undergo transformations through physical interactions, and these transformations lead to changes in the physical properties that are of value for piezotechnical applications.

The well-known classification of the two-component composites with planar interfaces was first put forward by Newnham et al. [64]. This classification is concerned with the so-called *connectivity* of each component. Connectivity is regarded as one of the main characteristics of the microstructure and expressed by the numbers of dimensions (or co-ordinate axes) in which each component is continuously distributed between limiting surfaces of the composite sample. The distribution of a self-connected state of a component can take place along zero, one, two, or three co-ordinate axes, i.e., connectivity $\alpha = 0, 1, 2$, or 3 for component 1 and connectivity $\beta = 0, 1, 2$, or 3 for component 2. The connectivity of a two-component composite is written in a general form [10, 64] of α–β where the connectivity of the piezoelectric or most piezo-active component takes the first position (α). In the case of $\alpha \le \beta$, the n-component composites are described by

$(n + 3)!/(3!\ n!)$ connectivities [64], and for instance, the number of connectivities is 10 for $n = 2$. It is also possible to introduce 10 alternative connectivities α–β at $\alpha \geq \beta$ and $n = 2$.

The concept of connectivity [64] is fundamental in developing an understanding of the electromechanical interaction between components within piezo-composites, in the study of the distribution of internal electric and mechanical fields and in the interpretation of experimental or calculated data related to composites with specific connectivity. The connectivity, of the piezo-active composites is crucial in influencing the piezoelectric response and electromechanical coupling of these materials [10, 26, 65, 66].

The entire complex of the α–β connectivity patterns, their evolution and interconnections between them in the two-component composites with planar interfaces were analysed in work [10]. To describe the evolution and determine the effective electromechanical properties in the α–β piezo-composite with planar interfaces, a group of so-called *junction* connectivity patterns (1–1, 1–3, 2–2, and 3–1) was introduced. Knowledge of the evolution of the connectivity patterns enables the analysis of the α–β piezo-composites and their effective electromechanical properties [10].

The problem of predicting the effective physical properties in piezo-active composites is of interest to many specialists undertaking theoretical and experimental studies of these heterogeneous materials. A formulation of the related problem on the effective electromechanical properties in a piezo-active composite is given in monograph [5]. If the averaged components of the electric field $\langle E_f \rangle$ and the mechanical strain $\langle \xi_q \rangle$ in a macroscopic region of the composite sample are independent of co-ordinates x_j(j = 1, 2 and 3), then (1.4) and (1.5) for the piezoelectric medium can be written for the piezo-active composite in the following form:

$$\langle \sigma_p \rangle = c_{pq}^{*E} \langle \xi_q \rangle - e_{fp}^{*} \langle E_f \rangle \tag{1.33}$$

$$\langle D_k \rangle = e_{kl}^{*} \langle \xi_l \rangle + \varepsilon_{kr}^{*\xi} \langle E_r \rangle \tag{1.34}$$

In (1.33) and (1.34) elastic moduli c_{pq}^{*E}, piezoelectric coefficients e_{fp}^{*} and dielectric permittivities $\varepsilon_{kr}^{*\xi}$ constitute the full set of effective electromechanical constants[1] that are found [5] by taking into account equations of electric and mechanical equilibrium and boundary conditions at the surface of the macroscopic region in the heterogeneous medium. The determination of the effective electromechanical properties of the piezo-active composite is concerned with an averaging of a series of vector and tensor components of the electric and mechanical fields, for example, ξ_q, E_f, σ_p, and D_k. This averaging is performed on

[1]Hereafter we use asterisk (*) to denote the effective properties and related parameters of the composite.

volume fractions of the composite components ($n = 1, 2,\ldots$), for which the full sets of electromechanical constants, e.g. $c_{pq}^{(n),E}$, $e_{fp}^{(n)}$ and $\varepsilon_{kr}^{(n),\xi}$ are known.

Taking into consideration volume-fraction dependences of the effective electromechanical properties of a composite [5, 10] with a specific microgeometry [66], one can predict the behaviour of the ECFs, figures of merit and other parameters that are of interest for energy-harvesting applications [46, 67–69].

The next step in studying the piezo-composites is concerned with the orientations effects [26]. The orientation effects broaden our outlook in the field of *composition—structure—properties* relations [10, 12]. Examination of the orientation effects can promote an improvement of the performance of the piezo-composites, the anisotropy of their piezoelectric coefficients and ECFs, the hydrostatic piezoelectric response, etc. Fundamental links between the domain orientations and electromechanical properties in relaxor-ferroelectric SCs with high piezoelectric activity represent a considerable opportunity for improving the piezoelectric performance of the SC/polymer and SC/porous polymer composites. In comparison to the composites based on relaxor-ferroelectric SCs, the orientation effects in the conventional FC-based composites are less pronounced [26], but should be taken into account when predicting the piezoelectric properties and related parameters.

1.6 Conclusion

This chapter has been devoted to the introduction and description of the electromechanical properties and related characteristics of piezoelectric media and the consideration of the piezoelectric performance of modern piezoelectric materials. In the last decades the most important piezoelectric materials have been poled FCs, relaxor-ferroelectric, SCs with engineered domain structures and piezo-active composites with at least one of the aforementioned components. The use of highly-effective relaxor-ferroelectric SCs with compositions near the morphotropic phase boundary is of significant interest due to their electromechanical properties compared to those of the conventional FCs. The excellent electromechanical properties of the relaxor-ferroelectric PMN–*x*PT and PZN–*y*PT SCs (Table 1.1) in the domain-engineered/heterophase states are a result of their large values of piezoelectric coefficients d_{ij}, elastic compliances s_{ab}^{E}, ECFs k_{ij}, and squared figures of merit $(Q_{3j})^2$. The ECFs k_{ij} and squared figures of merit $(Q_{3j})^2$ are parameters that are specifically related to the energy-harvesting characteristics of the piezoelectric material, and these characteristics strongly depend on its piezoelectric activity.

The electromechanical properties and related characteristics of the poled FCs (Table 1.2) vary in a wide range as a result of the strong dependence of these properties on microstructure, composition, poling conditions and technological factors such as a range of sintering temperatures, hot-pressure parameters, etc. The lower piezoelectric activity of the poled FCs in comparison to the aforementioned relaxor-ferroelectric SCs leads to smaller values of $|k_{ij}|$ and $(Q_{3j})^2$ (Table 1.3), and this interconnection is to be taken into account when selecting potential piezoelectric energy-harvesting materials.

In general the electromechanical properties of the domain-engineered SCs, poled FC and piezo-active composite materials are regarded as 'effective properties' in accordance with features of the microstructure, domain structure, arrangement of components and connectivity. The piezo-active composites based on ferroelectrics are the final link in the hierarchy-of-properties chains of 'single-domain SC → polydomain SC → FC → composite' and 'single-domain SC → polydomain SC → heterophase SC → composite', and this makes the piezo-active composites a unique range of modern functional materials whose effective properties can be tailored in wide ranges due to a range of factors (physical, chemical, microgeometric, technological, etc.) Variations of the effective electromechanical properties and related parameters in the piezo-active composites open up new possibilities for piezoelectric energy-harvesting applications of these materials.

References

1. Levanyuk AP, Sannikov DG (1994) Piezoelectrics. In: Prokhorov AM (ed) Physics encyclopaedia. Bolshaya Rossiyskaya Entsiklopedia, vol 4. Moscow (in Russian), pp 188–189
2. Ikeda T (1990) Fundamentals of piezoelectricity. Oxford University Press, Oxford
3. Zheludev IS (1971) Physics of crystalline dielectrics. Electrical properties, vol. 2. Plenum, New York
4. Uchino K (1997) Piezoelectric actuators and ultrasonic motors. Kluwer, Boston
5. Khoroshun LP, Maslov BP, Leshchenko PV (1989) Prediction of effective properties of piezo-active composite materials. Naukova Dumka, Kiev (in Russian)
6. Turik AV (1970) Elastic, piezoelectric, and dielectric properties of single crystals of $BaTiO_3$ with a laminar domain structure. Soviet Phys—Solid State 12:688–693
7. Aleshin VI (1990) Domain-orientation contribution into constants of ferroelectric polydomain single crystal. Zh Tekh Fiz 60:179–183 (in Russian)
8. Topolov VYu (2003) Domain wall displacements and piezoelectric activity of $KNbO_3$ single crystals. J Phys: Condens Matter 15:561–565
9. Gorish AV, Dudkevich VP, Kupriyanov MF, Panich AE, Turik AV (1999) Piezoelectric device-making. Physics of ferroelectric ceramics, vol. 1. Radiotekhnika, Moscow (in Russian)
10. Topolov VYu, Bowen CR (2009) Electromechanical properties in composites based on ferroelectrics. Springer, London
11. Siffert P (2008) Foreword. In: Heywang W, Lubitz K, Wersing W (eds) Piezoelectricity. Evolution and future of a technology. Springer, Berlin, p V
12. Newnham RE (2005) Properties of materials. Anisotropy, symmetry, structure. Oxford University Press, New York
13. Berlincourt DA, Cerran DR, Jaffe H (1964) Piezoelectric and piezomagnetic materials and their function in transducers. In: Mason W (ed) Physical acoustics. Principles and methods. Methods and devices. vol 1, Pt A. Academic Press, New York, pp 169–270
14. Kim H, Tadesse Y, Priya S (2009) Piezoelectric energy harvesting. In: Priya S, Inman DJ (eds) Energy harvesting technologies. Springer, New York, pp 3–39

15. Ayed SB, Abdelkefi A, Najar F, Hajj MR (2014) Design and performance of variable-shaped piezoelectric energy harvesters. J Intell Mater Syst Struct 25:174–186
16. Tang G, Yang B, Liu J-Q, Xu B, Zhu H-Y, Yang C-S (2014) Development of high performance piezoelectric d_{33} mode MEMS vibration energy harvester based on PMN-PT single crystal thick film. Sensors Actuators A 205:150–155
17. Chae M-S, Koh J-H (2014) Piezoelectric behavior of $(1 - x)(PbMgNbO_3–PbZrTiO_3) - x(BaTiO_3)$ ceramics for energy harvester applications. Ceram Int 40:2551–2555
18. Maiwa H, Sakamoto W (2013) Vibrational energy harvesting using a unimorph with PZT- or BT-based ceramics. Ferroelectrics 446:67–77
19. Uchino K, Ishii T (2010) Energy flow analysis in piezoelectric energy harvesting systems. Ferroelectrics 400:305–320
20. Yoon M-S, Mahmud I, Ur S-C (2013) Phase-formation, microstructure, and piezoelectric/dielectric properties of $BiYO_3$-doped $Pb(Zr_{0.53}Ti_{0.47})O_3$ for piezoelectric energy harvesting devices. Ceram Int 39:8581–8588
21. Gusarova E, Gusarov B, Zakharov D, Bousquet M, Viala B, Cugat O, Delamare J, Gimeno L (2013) An improved method for piezoelectric characterization of polymers for energy harvesting applications. J Phys: Conf Ser 476:012061
22. Yan Y, Cho K-H, Maurya D, Kumar A, Kalinin S, Khachaturyan A, Priya S (2013) Giant energy density in [001]-textured $Pb(Mg_{1/3}Nb_{2/3})O_3 - PbZrO_3 - PbTiO_3$ piezoelectric ceramics. Appl Phys Lett 102:042903
23. Islam RA, Priya S (2006) Realization of high-energy density polycrystalline piezoelectric ceramics. Appl Phys Lett 88:032903
24. Priya S (2010) Criterion for material selection in design of bulk piezoelectric energy harvesters. IEEE Trans Ultrason Ferroelectr Freq Control 57:2610–2612
25. Grekov AA, Kramarov SO, Kuprienko AA (1989) Effective properties of a transversely isotropic piezoelectric composite with cylindrical inclusions. Mech Compos Mater 25:54–61
26. Topolov VYu, Bisegna P, Bowen CR (2014) Piezo-active composites. Orientation effects and anisotropy factors. Springer, Berlin Heidelberg
27. Cross LE (2008) Relaxor ferroelectrics. In: Heywang W, Lubitz K, Wersing W (eds) Piezoelectricity. Evolution and future of a technology. Springer, Berlin, pp 131–156
28. Park S-E, Hackenberger W (2002) High performance single crystal piezoelectrics: applications and issues. Curr Opin Solid State Mater Sci 6:11–18
29. Smolensky GA, Bokov VA, Isupov VA, Krainik NN, Pasynkov RE, Sokolov AI, Yushin NK (1985) Physics of ferroelectric phenomena. Nauka, Leningrad (in Russian)
30. Noheda B (2002) Structure and high-piezoelectricity in lead oxide solid solutions. Curr Opin Solid State Mater Sci 6:27–34
31. Park S-E, Shrout TR (1997) Ultrahigh strain and piezoelectric behavior in relaxor based ferroelectric single crystals. J Appl Phys 82:1804–1811
32. Park S-E, Shrout TR (1997) Relaxor based ferroelectric single crystals for electro-mechanical actuators. Mater Res Innovations 1:20–25
33. Topolov VYu (2012) Heterogeneous ferroelectric solid solutions. Phases and Domain States. Springer, Berlin
34. Dammak H, Renault A-É, Gaucher P, Thi MP, Calvarin G (2003) Origin of the giant piezoelectric properties in the [001] domain engineered relaxor single crystals. Japan J Appl Phys 1(42):6477–6482
35. Topolov VYu, Turik AV (2002) An intermediate monoclinic phase and electromechanical interactions in $xPbTiO_3 - (1 - x)Pb(Zn_{1/3}Nb_{2/3})O_3$ crystals. Phys Solid State 44:1355–1362
36. Fu H, Cohen RE (2000) Polarization rotation mechanism for ultrahigh electromechanical response in single-crystal piezoelectrics. Nature 403:281–283
37. Davis M (2007) Picturing the elephant: Giant piezoelectric activity and the monoclinic phases of relaxor-ferroelectric single crystals. J Electroceram 19:23–45
38. Noheda B, Cox DE, Shirane G, Park S-E, Cross LE, Zhong Z (2001) Polarization rotation via a monoclinic phase in the piezoelectric 92 $\%PbZn_{1/3}Nb_{2/3}O_3$-8 $\%PbTiO_3$. Phys Rev Lett 86:3891–3894

39. Wada S, Tsurumi T (2002) Domain switching properties in PZN–PT single crystals with engineered domain configurations. Key Eng Mater 214–215:9–14
40. Fesenko EG, Gavrilyachenko VG, Semenchev AF (1990) Domain structure of multiaxial ferroelectric crystals. Rostov University Press, Rostov-on-Don (in Russian)
41. Liu T, Lynch CS (2003) Ferroelectric properties of [110], [001] and [111] poled relaxor single crystals: measurements and modeling. Acta Mater 51:407–416
42. Hong YK, Moon KS (2005) Single crystal piezoelectric transducers to harvest vibration energy. Proc SPIE Optomechatronic Actuators Manipulation 6048:60480E
43. Sun C, Qin L, Li F, Wang Q-M (2009) Piezoelectric energy harvesting using single crystal $Pb(Mg_{1/3}Nb_{2/3})O_3$–$x$$PbTiO_3$ (PMN-PT) device. J Intell Mater Syst Struct 20:559–568
44. Moon SE, LeeSQ Lee S-K, Lee Y-G, Yang YS, Park K-H, Kim J (2009) Sustainable vibration energy harvesting based on Zr-doped PMN-PT piezoelectric single crystal cantilevers. ETRI J 31:688–694
45. Song HJ, Choi YT, Wang G, Wereley NM (2009) Energy harvesting utilizing single crystal PMN-PT material and application to a self-powered accelerometer. J Mech Des 131:091008
46. Ren K, Liu Y, Geng X, Hofmann HF, Zhang QM (2006) Single crystal PMN-PT/epoxy 1–3 composite for energy-harvesting application. IEEE Trans Ultrason Ferroelectr Freq Control 53:631–638
47. Turik AV, Topolov VYu, Aleshin VI (2000) On a correlation between remanent polarization and piezoelectric coefficients of perovskite-type ferroelectric ceramics. J Phys D Appl Phys 33:738–743
48. Turik AV, Chernobabov AI (1977) On an orientation contribution in dielectric, piezoelectric and elastic constants of ferroelectric ceramics. Zh Tekh Fiz 47:1944–1948 (in Russian)
49. Aleshin VI (1991) Spherical inclusion in an anisotropic piezo-active medium. Kristallografiya 36:1352–1357 (in Russian)
50. Aleshin VI (1987) Properties of textures being formed on the basis of non-180° reorientations. Kristallografiya 32:422–426 (in Russian)
51. Bondarenko EI, Topolov VYu, Turik AV (1990) The effect of 90° domain wall displacements on piezoelectric and dielectric constants of perovskite ceramics. Ferroelectrics 110:53–56
52. Bondarenko EI, Topolov VYu, Turik AV (1991) The role of 90° domain wall displacements in forming physical properties of perovskite ferroelectric ceramics. Ferroelectr Lett Sect 13: 13–19
53. Topolov VYu, Bondarenko EI, Turik AV, Chernobabov AI (1993) The effect of domain structure on electromechanical properties of $PbTiO_3$-based ferroelectrics. Ferroelectrics 140:175–181
54. Turik AV, Topolov VYu (1997) Ferroelectric ceramics with a large piezoelectric anisotropy. J Phys D Appl Phys 30:1541–1549
55. Ruschmeyer K, Helke G, Koch J, Lubitz K, Möckl T, Petersen A, Riedel M, Schönecker A (1995) Piezokeramik: Grundlagen, Werkstoffe, Applikationen. Expert-Verlag, Renningen-Malmsheim
56. Algueró M, Alemany C, Pardo L, González AM (2004) Method for obtaining the full set of linear electric, mechanical and electromechanical coefficients and all related losses of a piezoelectric ceramic. J Am Ceram Soc 87:209–215
57. Dantsiger AYa, Razumovskaya ON, Reznitchenko LA, Grineva LD, Devlikanova RU, Dudkina SI, Gavrilyatchenko SV, Dergunova NV, Klevtsov AN (1994) Highly effective piezoceramic materials (Handbook). Kniga, Rostov-on-Don (in Russian)
58. Haertling G (1999) Ferroelectric ceramics: history and technology. J Am Ceram Soc 82: 797–818
59. Zhang R, Jiang B, Cao W (2001) Elastic, piezoelectric, and dielectric properties of multidomain $0.67Pb(Mg_{1/3}Nb_{2/3})O_3$–$0.33PbTiO_3$ single crystals. J Appl Phys 90:3471–3475
60. Zhang R, Jiang W, Jiang B, Cao W (2002) Elastic, dielectric and piezoelectric coefficients of domain engineered $0.70Pb(Mg_{1/3}Nb_{2/3})O_3$–$0.30PbTiO_3$ single crystal. In: Cohen RE (ed) Fundamental physics of ferroelectrics. Melville, American Institute of Physics, pp 188–197

61. Liu G, Jiang W, Zhu J, Cao W (2011) Electromechanical properties and anisotropy of single- and multi-domain $0.72Pb(Mg_{1/3}Nb_{2/3})O_3$–$0.28PbTiO_3$ single crystals. Appl Phys Lett 99:162901–162903
62. Yin J, Jiang B, Cao W (2000) Elastic, piezoelectric, and dielectric properties of $0.955Pb(Zn_{1/3}Nb_{2/3})O_3$–$0.045PbTiO_3$ single crystals. IEEE Trans Ultrason Ferroelectr Freq Control 47:285–291
63. Zhang R, Jiang B, Cao W, Amin A (2002) Complete set of material constants of $0.93Pb(Zn_{1/3}Nb_{2/3})O_3$–$0.07PbTiO_3$ domain engineered single crystal. J Mater Sci Lett 21: 1877–1879
64. Newnham RE, Skinner DP, Cross LE (1978) Connectivity and piezoelectric-pyroelectric composites. Mater Res Bull 13:525–536
65. Gururaja TR, Safari A, Newnham RE, Cross LE (1988) Piezoelectric ceramic-polymer composites for transducer applications. In: Levinson LM (ed) Electronic ceramics: properties, devices, and applications. Marcel Dekker, New York, pp 92–128
66. Topolov VYu, Glushanin SV (2002) Evolution of connectivity patterns and links between interfaces and piezoelectric properties of two-component composites. J Phys D Appl Phys 35:2008–2014
67. Swallow LM, Luo JK, Siores E, Patel I, Dodds D (2008) A piezoelectric fibre composite based energy harvesting device for potential wearable applications. Smart Mater Struct 17:025017
68. Qi Y, McAlpine MC (2010) Nanotechnology-enabled flexible and biocompatible energy harvesting. Energy Environ Sci 3:1275–1285
69. Guyomar D, Lallart M (2011) Recent progress in piezoelectric conversion and energy harvesting using nonlinear electronic interfaces and issues in small scale implementation. Micromachines 2:274–294
70. Bechmann R (1956) Elastic, piezoelectric, and dielectric constants of polarized barium titanate ceramics and some applications of the piezoelectric equations. J Acoust Soc Am 28:347–350
71. Huang JH, Kuo W-S (1996) Micromechanics determination of the effective properties of piezoelectric composites containing spatially oriented short fibers. Acta Mater 44:4889–4898
72. Dunn ML, Taya M (1993) Electromechanical properties of porous piezoelectric ceramics. J Am Ceram Soc 76:1697–1706
73. Levassort F, Lethiecq M, Millar C, Pourcelot L (1998) Modeling of highly loaded 0–3 piezoelectric composites using a matrix method. IEEE Trans Ultrason Ferroelectr Freq Control 45:1497–1505
74. Levassort F, Lethiecq M, Certon D, Patat F (1997) A matrix method for modeling electroelastic moduli of 0–3 piezo-composites. IEEE Trans Ultrason Ferroelectr Freq Control 44:445–452
75. Jaffe B, Cook WR, Jaffe H (1971) Piezoelectric ceramics. Academic Press, London
76. Ikegami S, Ueda I, Nagata T (1971) Electromechanical properties of $PbTiO_3$ ceramics containing La and Mn. J Acoust Soc Am 50:1060–1066
77. Xu Y (1991) Ferroelectric materials and their applications. North-Holland, Amsterdam
78. Nagatsuma K, Ito Y, Jyomura S, Takeuchi H, Ashida S (1985) Elastic properties of modified $PbTiO_3$ ceramics with zero temperature coefficients. In: Taylor GW (ed) Ferroelectricity and related phenomena, vol 4. Piezoelectricity. Gordon and Breach Science Publishers, New York, pp 167–176
79. Levassort F, Thi MP, Hemery H, Marechal P, Tran-Huu-Hue L-P, Lethiecq M (2006) Piezoelectric textured ceramics: effective properties and application to ultrasonic transducers. Ultrasonics 441(Suppl. 1):e621–e626

Chapter 2
Electromechanical Coupling Factors and Their Anisotropy in Piezoelectric and Ferroelectric Materials

The ECF characterises the conversion of electrical energy into the mechanical form and the conversion of mechanical energy into the electric form (see work [1, 2] and Sect. 1.2). A system of the ECFs (see, for example, (1.21)–(1.27) for poled FCs) is introduced to describe the conversion and takes into account the symmetry of a piezoelectric material, orientations of its crystallographic axes, input and output arrangements, etc. The ECF is often defined for a respective oscillation mode in a piezoelectric element at the static limit of frequency $\omega \to 0$. It should be added that the concept of the electromechanical coupling in piezoelectrics originates from a factor that indicates the degree of the inductive coupling in a transformer [1]. A larger absolute value of the EFC $|k|$ implies a more efficient performance of a piezoelectric material at a specific mode of oscillation and k^2 is used to characterise the magnitude of the transducer bandwidth [1]. Knowledge of the ECF values of piezoelectrics is important for energy-harvesting applications in the context of relationships between the amount of stored and supplied energy. These relationships depend on the electromechanical properties and other characteristics.

In Chap. 2 we consider and compare the electromechanical properties and ECFs of various SC, ceramic and composite materials to show their advantages in piezotechnical applications. Of specific interest are advantages of modern ferroelectric materials over conventional piezoelectrics.

2.1 Piezoelectric Single Crystals

Quartz (SiO_2) has been the first and widespread piezoelectric material [1–4] after the discovery of the piezoelectric effects by brothers P. Curie and J. Curie in 1880. Below $T = 847$ K quartz SC is related to the 32 symmetry class, and the matrix of the piezoelectric coefficients d_{ij} is given by

C.R. Bowen et al., *Modern Piezoelectric Energy-Harvesting Materials*,
Springer Series in Materials Science 238, DOI 10.1007/978-3-319-29143-7_2

$$\|d\| = \begin{pmatrix} d_{11} & -d_{11} & 0 & d_{14} & 0 & 0 \\ 0 & 0 & 0 & 0 & -d_{14} & -2d_{11} \\ 0 & 0 & 0 & 0 & 0 & 0 \end{pmatrix}$$

According to room-temperature experimental data [4], these piezoelectric coefficients are relatively small ($d_{11} = 2.30$ pC/N and $d_{14} = -0.693$ pC/N) in comparison to d_{ij} of various ferroelectric SCs and FCs [2–7]. Taking into account elastic compliances of the quartz SC $s_{11}^E = 12.77 \cdot 10^{-12}\,\mathrm{Pa}^{-1}$ and $s_{44}^E = 20.02 \cdot 10^{-12}\,\mathrm{Pa}^{-1}$ and its dielectric permittivity $\varepsilon_{11}^\sigma / e_0 = 4.520$ [4], we obtain the ECFs as follows:

$$k_{11} = d_{11} / \left(\varepsilon_{11}^\sigma s_{11}^E\right)^{-1/2} = 0.102 \text{ and } k_{14} = d_{14} / \left(\varepsilon_{11}^\sigma s_{44}^E\right)^{-1/2} = 0.0245 \qquad (2.1)$$

The k_{1j} values from (2.1) are to be compared to k_{ij} of SCs from different symmetry classes (see, for instance, data in Table 2.1 and Sect. 2.2).

To the best of our knowledge, the important group of piezoelectric SCs consist of compounds such as α-ZnS, CdS, CdSe, ZnO, etc. [4] from the 6*mm* symmetry class. The electromechanical properties of these SCs are shown in Table 2.1, the ECFs for the 6*mm* symmetry class have the form of k_{ij} from (1.21)–(1.23). An anisotropy factor

$$\zeta_{k3j} = k_{33} / k_{31} \qquad (2.2)$$

is related to the anisotropy of electromechanical properties of the piezoelectric material. The anisotropy factor ζ_{k3j} from (2.2) for the 6*mm* symmetry class is represented as

$$\zeta_{k3j} = (d_{33}/d_{31})\left(s_{11}^E / s_{33}^E\right)^{1/2} \qquad (2.3)$$

By analogy with (2.3) we use the anisotropy factor

$$\zeta_{d3j} = d_{33} / d_{31} \qquad (2.4)$$

to describe the anisotropy of the piezoelectric coefficients d_{3j}. Knowledge of the ECFs from (1.21)–(1.26) and their anisotropy factors from (2.2) and (2.3) are of benefit for design of efficient energy-harvesting materials since we determine directions that are preferable for the conversion and propagation of energy.

We remind the reader that the matrix of the piezoelectric coefficients d_{ij} in the 6*mm* symmetry class and ∞*mm* symmetry group has the form [1–4]

$$\|d\| = \begin{pmatrix} 0 & 0 & 0 & 0 & d_{15} & 0 \\ 0 & 0 & 0 & d_{15} & 0 & 0 \\ d_{31} & d_{31} & d_{33} & 0 & 0 & 0 \end{pmatrix} \qquad (2.5)$$

Table 2.1 Experimental values of elastic compliances s_{ab}^E (in 10^{-12} Pa^{-1}), piezoelectric coefficients d_{ij} (in pC/N), dielectric permittivities $\varepsilon_{pp}^{\sigma}$ [4], ECFs k_{ij}, and anisotropy factor ζ_{k3j} of piezoelectric SCs (4*mm* symmetry) at room temperature

	α-ZnS	CdS	CdSe	ZnO
s_{11}^E	11.12	20.69	23.38	7.858
s_{12}^E	–4.56	–9.93	–11.22	–3.432
s_{13}^E	–1.40	–5.81	–5.72	–2.206
s_{33}^E	8.47	16.97	17.35	6.94
s_{44}^E	34.4	66.49	75.95	23.57
s_{66}^E	31.4	61.36	–	22.58
d_{31}	–1.1	–5.18	–3.92	–5.2
d_{33}	3.2	10.32	7.84	10.6
d_{15}	–2.6	–13.98	–10.51	–13.9
$\varepsilon_{11}^{\sigma}/\varepsilon_0$	8.65	9.35	9.70	9.26
$\varepsilon_{33}^{\sigma}/\varepsilon_0$	8.70	10.38	10.6	11.0
k_{31}	–0.039	–0.12	–0.084	–0.19
k_{33}	0.13	0.26	0.19	0.41
k_{15}	–0.051	–0.070	–0.13	–0.32
ζ_{k3j}	–3.3	–2.2	–2.3	–2.2

As follows from Table 2.1, the piezoelectric activity of the listed SCs is higher than the piezoelectric activity of quartz. As a consequence, larger values of $|k_{ij}|$ are achieved in comparison to $|k_{1j}|$ from (2.1). At the same time, absolute values of $|\zeta_{k3j}|$ remain relatively small in comparison to $|\zeta_{k3j}|$ for conventional PZT-type FCs (see, for instance, [3, 5]).

2.2 Ferroelectric Single Crystals

2.2.1 Single-Domain States

Examples of full sets of electromechanical constants of the perovskite-type ferroelectric SCs in the single-domain state are given in Table 2.2. The matrix of the piezoelectric coefficients d_{ij} in the 4*mm* symmetry class coincides with that given in (2.5). It is assumed that the spontaneous polarisation vector $\boldsymbol{P}_s$ of the single-domain SC is parallel to the co-ordinate axis OX_3 that is parallel to the crystallographic axis z of the perovskite unit cell.

It is seen that the two sets of electromechanical constants of the single-domain $BaTiO_3$ SC (Table 2.2) lead to differences between ECFs, however their anisotropy remains moderate. From the set of electromechanical constants of the single-domain $PbTiO_3$ SC it can be seen that while there is some piezoelectric anisotropy, $\zeta_{d3j} = -5.32$, the ECFs are characterised by a lower degree of anisotropy, so that

Table 2.2 Experimental values of elastic compliances s_{ab}^E (in 10^{-12} Pa^{-1}), piezoelectric coefficients d_{ij} (in pC/N), dielectric permittivities $\varepsilon_{pp}^{\sigma}$, ECFs k_{ij}, and anisotropy factor ζ_{k3j} of single-domain $BaTiO_3$ and $PbTiO_3$ SCs (4*mm* symmetry) at room temperature

	$BaTiO_3^a$	$BaTiO_3^b$	$PbTiO_3^b$
s_{11}^E	8.05	8.20	6.7
s_{12}^E	−2.35	−2.46	−2.00
s_{13}^E	−5.24	−5.66	−6.30
s_{33}^E	15.7	16.20	34.80
s_{44}^E	18.4	15.40	12.90
s_{66}^E	8.84	9.32	7.60
d_{31}	−34.7	−38	−26
d_{33}	85.7	110	160
d_{15}	587	407	56
$\varepsilon_{11}^{\sigma}/\varepsilon_0$	4100	3200	140
$\varepsilon_{33}^{\sigma}/\varepsilon_0$	160	200	115
k_{31}	−0.325	−0.315	−0.315
k_{33}	0.575	0.650	0.850
k_{15}	0.718	0.616	0.443
ζ_{k3j}	−1.77	−2.06	−2.70

[a]Electromechanical constants from [4]
[b]Electromechanical constants from [7]

$\zeta_{k3j} \approx \zeta_{d3j}/2$. The difference between the anisotropy factors ζ_{d3j} and ζ_{k3j} is a result of a large difference between the elastic compliances s_{33}^E and s_{11}^E (see the 4th column in Table 2.2) in the single-domain state.

2.2.2 *Polydomain States*

An important example of the electromechanical properties and ECFs of the polydomain $BaTiO_3$ SC is shown in Table 2.3. In this case we consider a SC sample split into 90° domains (Fig. 2.1) with the conventional 'head-to-tail' arrangement [7, 8]. Spontaneous polarisation vectors of these domain types are $\boldsymbol{P}_s^{(1)}$ $(-P; 0; P)$ and $\boldsymbol{P}_s^{(2)}(P; 0; P)$, and the planar 90° domain walls are parallel to the (X_1OX_2) plane of the rectangular co-ordinate system $(X_1X_2X_3)$. The 90° domain structure is regarded as regular, and at equal volume fractions of the domain types shown ($m = 1$ to $m = 1/2$), the average spontaneous polarisation vector is $\boldsymbol{P}_s \uparrow\uparrow OX_3$. The co-ordinate axes OX_1, OX_2 and OX_3 are parallel to the main crystallographic axes x, y and z, respectively. The electromechanical constants listed in Table 2.3 were calculated [8] assuming that the 90° domain walls are motionless. A contribution from the domain-wall displacements in the electromechanical properties of the SC split into the 90° domains was evaluated in work [9].

Table 2.3 Caluclated values[a] of elastic compliances s_{ab}^E (in 10^{-12} Pa^{-1}), piezoelectric coefficients d_{ij} (in pC/N), dielectric permittivities ε_{pp}^σ [8], ECFs k_{ij}, and anisotropy factor ζ_{k3j} of the polydomain $BaTiO_3$ SC (*mm*2 symmetry) at room temperature

	Polydomain $BaTiO_3$
s_{11}^E	7.92
s_{12}^E	−3.80
s_{13}^E	−1.28
s_{22}^E	8.05
s_{23}^E	−3.80
s_{33}^E	7.92
s_{44}^E	11.9
s_{55}^E	30.2
s_{66}^E	13.6
d_{31}	−189
d_{32}	−24.5
d_{33}	225
d_{15}	126
d_{24}	269
$\varepsilon_{11}^\sigma/\varepsilon_0$	265
$\varepsilon_{22}^\sigma/\varepsilon_0$	2680
$\varepsilon_{33}^\sigma/\varepsilon_0$	2130
k_{31}	−0.489
k_{32}	−0.0629
k_{33}	0.582
k_{15}	0.473
k_{24}	0.506
ζ_{k31}	−1.19
ζ_{k32}	−9.25

[a]Calculations were performed [8] using the full set of electromechanical constants of the single-domain $BaTiO_3$ SC (see the 2nd column of Table 2.2)

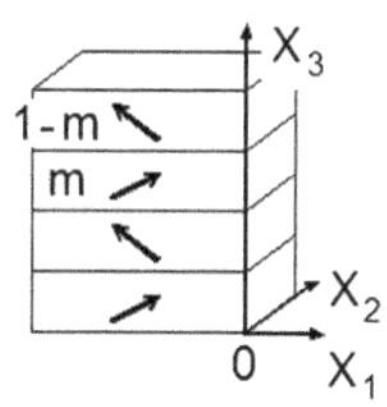

Fig. 2.1 Schematic arrangement of 90° domains in the tetragonal ferroelectric phase of the $BaTiO_3$-type SC. $(X_1X_2X_3)$ is the rectangular co-ordinate system. Spontaneous polarisation vectors of domains are shown by *arrows*, *m* and 1 − *m* are volume fractions of these domains

Our comparison of data on the single-domain and polydomain SCs (see the 2nd columns of Tables 2.2 and 2.3) enables us to underline the pronounced anisotropy of d_{3j} and k_{3j} in the polydomain state: for example, $\zeta_{d32} = d_{33}/d_{32} = -9.18$ and $\zeta_{k32} = k_{33}/k_{32} = -9.25$. This feature is a consequence of the arrangement of the spontaneous polarisation vectors $\boldsymbol{P}_s^{(1)}$ and $\boldsymbol{P}_s^{(2)}$ in the (X_1OX_3) plane that is perpendicular to the 90° domain wall (see Fig. 2.1).

2.3 Domain-Engineered Relaxor-Ferroelectric Single Crystals with High Piezoelectric Activity

The perovskite-type solid solutions of $(1 - x)Pb(B_1,B_2)O_3 - xPbTiO_3$ combine the physical properties of the relaxor-type and 'normal' ferroelectric components. Excellent electromechanical properties are observed near the morphotropic phase boundary in such solid solutions as PMN–*x*PT and PZN–*y*PT with engineered non-180° domain structures (see, for instance, data in Table 1.1). We remind the reader that domain-engineered PMN–*x*PT and PZN–*y*PT SCs poled along specific crystallographic directions (often along [001], [011] or [111] of the perovskite unit cell) exhibit very high piezoelectric activity and significant electromechanical coupling [10–13], which are important for energy-harvesting applications.

Our evaluations of ECFs (Table 2.4) show that absolute values of k_{3j} are generally larger than those related to the polydomain $BaTiO_3$ SC (Table 2.3). We note that values of $0.91 < k_{33} < 0.96$ and $k_{33} \approx |k_p|$ are achieved in various PMN–*x*PT and PZN–*y*PT SCs (Table 2.4) near the morpotropic phase boundary due to the very high piezoelectric activity and considerable elastic compliance of these domain-engineered SCs. Relatively large values of the hydrostatic ECF ($0.15 < k_h < 0.17$) are achieved in some compositions. At the same time the anisotropy of d_{3j} and k_{3j} in the domain-engineered PMN–*x*PT and PZN–*y*PT SCs is not large. These and related SCs are of interest as highly piezo-active components in modern composites [6] wherein orientation effects play an important role in the formation of the piezoelectric performance and hydrostatic response.

Table 2.4 ECFs of [001]-poled domain-engineered PMN–*x*PT and PZN–*y*PT SCs (4*mm* symmetry) at room temperature

SC	k_{31}	k_{33}	k_{15}	k_h	k_t	k_p
PMN–0.33PT	–0.594	0.957	0.322	0.167	0.691	–0.918
PMN–0.30PT	–0.486	0.916	0.284	0.157	0.651	–0.850
PMN–0.28PT	–0.387	0.915	0.257	0.0629	0.647	–0.923
PZN–0.045PT	–0.499	0.897	0.214	0.0843	0.441	–0.874
PZN–0.07PT	–0.582	0.924	0.271	0.0680	0.475	–0.901
PZN–0.08PT	–0.598	0.932	0.248	–0.0258	0.443	–0.917

Calculations were performed using experimental data from Table 1.1

Table 2.5 ECFs of perovskite-type FCs[a] at room temperature

FC	k_{31}	k_{33}	k_{15}	k_h	k_t	k_p
$BaTiO_3$ (I)	–0.208	0.493	0.467	0.0830	0.382	–0.354
$BaTiO_3$ (II)	–0.211	0.503	0.481	0.0847	0.397	–0.355
$(Ba_{0.917}Ca_{0.083})TiO_3$	–0.186	0.458	0.479	0.0899	0.363	–0.315
TBK-3	–0.167	0.412	0.429	0.0714	0.329	–0.280
TBKS	–0.123	0.398	0.324	0.130	0.351	–0.203
NBS-1	–0.217	0.551	0.422	0.126	0.463	–0.359
ZTS-19	–0.281	0.646	0.626	0.127	0.495	–0.506
ZTS-24	–0.309	0.643	0.666	0.103	0.436	–0.539
ZTSNV-1	–0.292	0.588	0.703	0.106	0.524	–0.521
PZT-4	–0.332	0.692	0.703	0.123	0.508	–0.573
PZT-5	–0.343	0.703	0.684	0.0801	0.488	–0.601
PZT-5H	–0.262	0.623	0.593	0.110	0.513	–0.445
PZT-7A	–0.299	0.659	0.671	0.155	0.505	–0.506
Navy Type VI	–0.373	0.723	0.701	0.0753	0.485	–0.627
PZ 27	–0.338	0.706	0.616	0.185	0.469	–0.604
PZ 34	–0.0339	0.394	0.231	0.402	0.464	–0.0536
PCR-1, hp[b]	–0.354	0.727	0.674	0.133	0.513	–0.622
PCR-7, hp	–0.384	0.833	0.720	0.0877	0.658	–0.711
PCR-7M, hp	–0.398	0.816	0.708	0.0921	0.595	–0.716
PCR-8, hp	–0.330	0.660	0.624	0.0828	0.446	–0.588
PCR-8, ct	–0.320	0.686	0.674	0.0750	0.540	–0.565
PCR-13, ct	–0.243	0.501	0.427	0.0298	0.418	–0.427
PCR-21, hp	–0.290	0.644	0.520	0.0860	0.493	–0.522
PCR-63, hp	–0.188	0.439	0.367	0.0577	0.337	–0.332
PCR-73, hp	–0.390	0.770	0.723	0.161	0.469	–0.700
$Pb(Zr_{0.54}Ti_{0.46})O_3$	0.280	0.626	0.735	0.148	0.481	–0.469
$Pb(Zr_{0.52}Ti_{0.48})O_3$	–0.313	0.671	0.696	0.123	0.512	–0.527
$(Pb_{0.94}Sr_{0.06})(Ti_{0.47}Zr_{0.53})O_3$	–0.327	0.684	0.695	0.122	0.501	–0.545
PMN– 0.35PT	–0.215	0.414	0.802	0.00785	0.232	–0.0846
Modified $PbTiO_3$ (I)	–0.0406	0.456	0.279	0.270	0.454	–0.0642
Modified $PbTiO_3$ (II)	–0.0598	0.477	0.338	0.263	0.470	–0.0957
$(Pb_{0.9625}La_{0.025})(Ti_{0.99}Mn_{0.01})O_3$	–0.0401	0.443	0.300	0.288	0.453	–0.0633
$(Pb_{0.85}Nd_{0.10})(Ti_{0.99}Mn_{0.01})O_3$	–0.0439	0.446	0.373	0.298	0.459	–0.0633
$(Pb_{0,855}Nd_{0,11})(Ti_{0,94}Mn_{0,02}In_{0,04})O_3$	–0.0525	0.458	0.407	0.296	0.296	–0.0702

[a]Full sets of electromechanical constants of the FCs are given in [6]

[b]FC samples of the PCR type have been manufactured using either the conventional technology (ct) or hot pressing (hp). PCR is the abbreviation for the group "piezoelectric ceramics from Rostov-on-Don" (Russia) [6, 14, 15]

2.4 Poled Ferroelectric Ceramics

Data on ECFs of the widespread perovskite-type FCs in the poled state (Table 2.5) allow us to divide the ceramic compositions into groups as follows:

(i) FCs with $k_{33} \geq 0.7$ and moderate anisotropy of k_{3j},
(ii) FCs with $0.5 < k_{33} < 0.7$ and moderate anisotropy of k_{3j}
and (iii) FCs with $k_{33} < 0.5$ and pronounced anisotropy of k_{3j}.

In these FC groups the piezoelectric coefficient d_{33} and ECF k_{33} are smaller than those related to the domain-engineered PMN–*x*PT and PZN–*y*PT SCs (see Tables 1.1, 1.2 and 2.4), however the anisotropy of d_{3j} and k_{3j} in FCs (see (2.2)–(2.4)) varies in relatively wide ranges. In the group of highly anisotropic FCs based on $PbTiO_3$ (see compositions from modified $PbTiO_3$ (I) to $(Pb_{0,855}Nd_{0,11})(Ti_{0,94}Mn_{0,02}In_{0,04})O_3$ in the bottom part of Table 2.5), inequalities $k_{33}/|k_{31}| \gg 1$ and $k_t/|k_p| \gg 1$ hold mainly due to the large anisotropy of d_{3j} and s^E_{ab} (see Table 1.2).

It is important to note that the hydrostatic ECF k_h also varies in a wide range (see Table 2.5), and this feature of the electromechanical coupling is concerned with a specific balance of elastic, piezoelectric and dielectric constants of the FC. Another interesting example of the electromechanical coupling is illustrated by experimental data [16] from Table 2.6. There is a correlation between the piezoelectric properties and electromechanical coupling in PMN–*x*PT FCs near the morphotropic phase boundary: maxima of three parameters are related to $x = 0.345$ or located near this molar-concentration point. We add that the planar ECF $k_p \sim d_{31}$, and $\max|d_{31}|$ is also expected near the morphotropic phase boundary. Comparing data on k_t of PMN–*x*PT SCs (Table 2.4) and PMN–*x*PT FCs (Table 2.6), it can be seen that $\max k_t$ of the FC samples is comparable to that shown in Table 2.4 for SCs with $0.28 \leq x \leq 0.33$, i.e., close to the morphotropic phase boundary. Our comparison of

Table 2.6 Correlation between $\max d_{33}$ (in pC/N), $\max k_t$ and $\max|k_p|$ in PMN–*x*PT FCs[a] at room temperature

x	d_{33}	k_t	$\lvert k_p \rvert$
0.28	320	0.51	0.38
0.30	450	0.55	0.41
0.31	570	0.57	0.44
0.32	590	0.62	0.43
0.33	640	0.61	0.43
0.34	670	0.62	0.42
0.345[b]	720	0.62	0.45
0.35	700	0.61	0.44
0.355	650	0.60	0.42
0.36	585	0.59	0.38
0.38	465	0.32	0.15
0.40	370	0.22	0.10

[a]Experimental data are taken from [16]
[b]At $x = 0.345$ $\max d_{33}$ and $\max|k_p|$ are achieved, and diffuse $\max k_t$ can be found at $0.33 < x < 0.35$

electromechanical constants of the SC and FC samples at $x = \text{const}$ suggests that the balance of elastic and piezoelectric constants influences $k_t/|k_p|$.

Clearly the larger values of d_{33} and k_t are achieved in the domain-engineered SC where the non-180° domain orientation [10–13] plays an important role in the piezoelectric properties.

2.5 Two-Component Composites Based on Ferroelectrics

In this section we consider examples of the anisotropic electromechanical coupling in two-component piezo-active composites. These materials are characterised by similar microgeometric elements. Of particular interest are composites with 1–3 connectivity, i.e., a system of long piezoelectric rods in a large matrix that is either piezoelectric or piezo-passive. This composite architecture means that at least one of the piezoelectric components is distributed continuously along a co-ordinate axis. An anisotropy of electromechanical properties of the composite components and the rods oriented parallel to the poling axis influence the anisotropy of the ECFs. While the 1–3 piezo-composite has been widespread in the past decades and manufactured by various methods [6, 17, 18], we compare the performance of this composite at various combinations of components.

2.5.1 Ceramic/Polymer

Since the 1980s, the FC/polymer composite is one of main heterogeneous piezoelectric materials that is manufactured for various piezotechnical applications due to remarkable electromechanical properties [17, 18]. It is assumed that the 1–3

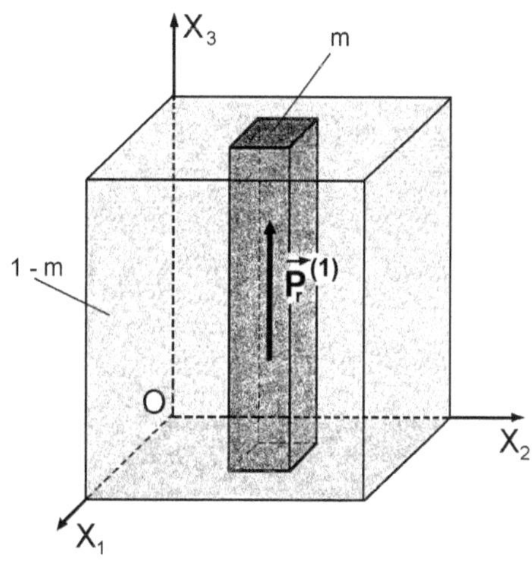

Fig. 2.2 Schematic of the 1–3 FC/polymer composite. $(X_1X_2X_3)$ is the rectangular co-ordinate system. m and $1 - m$ are volume fractions of FC and polymer, respectively

composite (Fig. 2.2) contains a system of FC rods in the form of rectangular parallelepipeds which are continuous in the OX_3 direction, having a square base and characterised by a regular square arrangement in the (X_1OX_2) plane. The FC rods are surrounded by a continuous polymer matrix and poled along the OX_3 direction, so that $\boldsymbol{P}_r^{(1)}$ is the remanent polarisation vector of each FC rod. We note that a square arrangement of the rods enables variation of their volume fraction m from 0 to 1.

The prediction of the effective electromechanical properties of the 1–3 composite (Fig. 2.2) with planar interfaces is carried out using the matrix approach [6, 17]. The averaging procedure implies averaging of the electromechanical constants of the FC rods and the polymer matrix in the OX_1 and OX_2 directions, in which the periodical structure of the composite is observed. In this procedure full sets of elastic compliances $s_{ij}^{(n),E}$ (measured at E = const), piezoelectric coefficients $d_{kl}^{(n)}$ and dielectric permittivities $\varepsilon_{ff}^{(n),\sigma}$ (measured at mechanical stress σ = const) of the components are used. We add that details of the matrix method applied to a 2–2 composite system with the laminar structure will be described in Sect. 3.2.1.

The effective electromechanical properties of the 1–3 composite (i.e., the full set of s_{ij}^{*E}, d_{kl}^{*} and $\varepsilon_{ff}^{*\sigma}$) are regarded as homogenised properties in the long-wave approximation [17], and s_{ij}^{*E}, d_{kl}^{*} and $\varepsilon_{ff}^{*\sigma}$ are represented as functions of the volume fraction m. Based on these properties, we analyse examples of a volume-fraction dependence of ECFs of the studied composite. The full sets of electromechanical constants of the FC and polymer components are taken from Tables 1.2 and 2.7, respectively. Following (1.25), (1.26), (2.2), and (2.3), we represent the anisotropy factors of ECFs in terms of effective electromechanical constants of the composite as a whole:

$$\zeta_{k3j}^{*} = k_{33}^{*}/k_{31}^{*} = \left(d_{33}^{*}/d_{31}^{*}\right)\left(s_{11}^{*E}/s_{33}^{*E}\right)^{1/2} \tag{2.6}$$

$$\zeta_{kt-p}^{*} = k_t^{*}/k_p^{*} = \left[e_{33}^{*}/\left(c_{33}^{*D}\varepsilon_{33}^{*\xi}\right)^{1/2}\right]/\left[k_{31}^{*}\left(2/\left(1-\sigma_E^{*}\right)\right)^{1/2}\right] \tag{2.7}$$

The hydrostatic ECF is written in the form analogous to that in (1.27):

$$k_h^{*} = d_h^{*}/\left(\varepsilon_{33}^{*\sigma}s_h^{*E}\right)^{1/2} \tag{2.8}$$

It is assumed that the effective electromechanical constants denoted by asterisks in (2.6)–(2.8) are functions of m.

Table 2.7 Elastic compliances $s_{ab}^{(n)}$ (in 10^{-12} Pa^{-1}) and dielectric permittivity $\varepsilon_{pp}^{(n)}$ of piezo-passive polymers at room temperature [6]

Polymer	$s_{11}^{(n)}$	$s_{12}^{(n)}$	$\varepsilon_{pp}^{(n)}/\varepsilon_0$
Araldite	216	–78	4.0
Polyurethane	405	–151	3.5
Elastomer	3300	–1480	5.0
Auxetic polyethylene	5260	4360	2.3

Table 2.8 Features of the volume-fraction dependence of ECFs of 1–3 FC/polymer composites based on modified $PbTiO_3$ FC (I)

k_{31}^*	k_{33}^*	ζ_{k3j}^*	k_t^*	k_p^*	$\zeta_{t\text{-}p}^*$	k_h^*
Modified $PbTiO_3$ FC (I)/araldite composite						
mink_{31}^* = –0.0886 (m = 0.076), maxk_{31}^* = –0.0386 (m = 0.909)	–[a]	minζ_{k3j}^* = –11.8 (m = 0.915)	–[a]	mink_p^* = –0.152 (m = 0.072), maxk_p^* = –0.0590 (m = 0.896)	min$\zeta_{t\text{-}p}^*$ = –7.66 (m = 0.900)	–[a]
Modified $PbTiO_3$ FC (I)/polyurethane composite						
mink_{31}^* = –0.0817 (m = 0.049), maxk_{31}^* = –0.0331 (m = 0.849)	–[a]	minζ_{k3j}^* = –13.7 (m = 0.852)	maxk_t^* = 0.454 (m = 0.993)	mink_p^* = –0.0497 (m = 0.851)	min$\zeta_{t\text{-}p}^*$ = –9.10 (m = 0.854)	–[a]
Modified $PbTiO_3$ FC (I)/elastomer composite						
mink_{31}^* = –0.0603 (m = 0.023), maxk_{31}^* = –0.0211 (m = 0.814)	–[a]	minζ_{k3j}^* = –21.6 (m = 0.834)	maxk_t^* = 0.454 (m = 0.963)	mink_p^* = –0.0227 (m = 0.846)	min$\zeta_{t\text{-}p}^*$ = –14.5 (m = 0.849)	–[a]
Modified $PbTiO_3$ FC (I)/auxetic polyethylene composite						
maxk_{31}^* = 0.124 (m = 0.010), k_{31}^* = 0 (m = 0.738)	–[a]	]–∞; ∞[	maxk_t^* = 0.455 (m = 0.939)	k_p^* = 0 (m = 0.738)	]–∞; ∞[	–[a]

[a]Monotonic increase

Values from Tables 2.8 and 2.9 suggest that the elastic anisotropy of the FC and polymer components plays an important role in the formation of the volume-fraction dependence of ECFs and their anisotropy. A more pronounced anisotropy of ECFs is observed in the presence of the modified $PbTiO_3$ FC and a softer polymer, while the larger values of longitudinal and thickness ECFs are achieved in the presence of PZT-5H FC which is a piezoelectric component with a relatively high piezoelectric activity. The presence of auxetic polymer (i.e., a medium with a negative Poisson's ratio [6]) leads to a considerable re-distribution of internal electric and mechanical fields and gives rise to changes in sign of some ECFs. As a result, the infinitely large anisotropy of ECFs ($|\zeta_{k3j}^*| \to \infty$ and $|\zeta_{kt\text{-}p}^*| \to \infty$) is observed irrespective of the FC component.

Examples of the volume-fraction behaviour of ECFs in the 1–3-type FC/auxetic polymer composite are shown in Fig. 2.3. The piezoelectric coefficient d_{31}^* (curve 1 in Fig. 2.3d) influences ECFs k_{31}^*, k_p^* and k_h^* (Fig. 2.3a–c). It is noteworthy that the auxetic polymer component promotes a wide range of volume fractions

Table 2.9 Features of the volume-fraction dependence of ECFs and anisotropy factors of 1–3 FC/polymer composites based on PZT-5H FC

k_{31}^*	k_{33}^*	ζ_{k3j}^*	k_t^*	k_p^*	$\zeta_{t\text{-}p}^*$	k_h^*
PZT-5H FC/araldite composite						
$\min k_{31}^* = -0.139$ ($m = 0.080$), $\max k_{31}^* = -0.133$ ($m = 0.273$)	–[a]	$\min \zeta_{k3j}^* = -4.50$ ($m = 0.329$)	$\max k_t^* = 0.584$ ($m = 0.620$)	$\min k_p^* = -0.244$ ($m = 0.073$)	$\min \zeta_{t\text{-}p}^* = -2.61$ ($m = 0.402$)	$\max k_h^* = 0.103$ ($m = 0.056$)
PZT-5H FC/polyurethane composite						
$\min k_{31}^* = -0.136$ ($m = 0.036$), $\max k_{31}^* = -0.109$ ($m = 0.320$)	–[a]	$\min \zeta_{k3j}^* = -5.60$ ($m = 0.342$)	$\max k_t^* = 0.599$ ($m = 0.600$)	$\min k_p^* = -0.247$ ($m = 0.036$)	$\min \zeta_{t\text{-}p}^* = -3.32$ ($m = 0.415$)	$\max k_h^* = 0.0978$ ($m = 0.030$)
PZT-5H FC/elastomer composite						
$\min k_{31}^* = -0.140$ ($m = 0.007$), $\max k_{31}^* = -0.0647$ ($m = 0.364$)	–[a]	$\min \zeta_{k3j}^* = -9.60$ ($m = 0.369$)	$\max k_t^* = 0.615$ ($m = 0.606$)	$\min k_p^* = -0.314$ ($m = 0.009$)	$\min \zeta_{t\text{-}p}^* = -5.66$ ($m = 0.483$)	$\max k_h^* = 0.0518$ ($m = 0.006$)
PZT-5H FC/auxetic polyethylene composite						
$\max k_{31}^* = 0.285$ ($m = 0.005$), $k_{31}^* = 0$ ($m = 0.386$)	–[a]	$]-\infty; \infty[$	$\max k_t^* = 0.621$ ($m = 0.593$)	$k_p^* = 0$ ($m = 0.386$)	$]-\infty; \infty[$	$\max k_h^* = 0.399$ ($m = 0.007$), $\min k_h^* = 0.0451$ ($m = 0.841$)

[a]Monotonic increase

($0.2 < m < 0.8$) where $k_p^* \approx 0$ (see curve 1 in Fig. 2.3b) and, therefore, large absolute values of the anisotropy factor ζ_{kt-p}^* are observed. As seen from curve 1 in Fig. 2.3a, the condition $k_{31}^* \approx 0$ is valid, however in a wider volume-fraction range. Such a comparison of curves of k_p^* and k_{31}^* provides an insight on the role of the elastic anisotropy of the composite in the formation of large anisotropy factors (2.6) and (2.7). We add that large absolute values of the anisotropy factors from (2.6) and (2.7) are beneficial for the piezoelectric energy-harvesting devices because of the preferential direction of the energy conversion at specific oscillation modes. In the cases of $k_{31}^* \approx 0$ or $k_p^* \approx 0$, the lateral oscillation mode does not play a role in the energy conversion, and vibration harvesting can take place along the poling axis of the composite material. At small volume fractions of FC m, there is a competition between the elastic properties of the components, and this competition influences

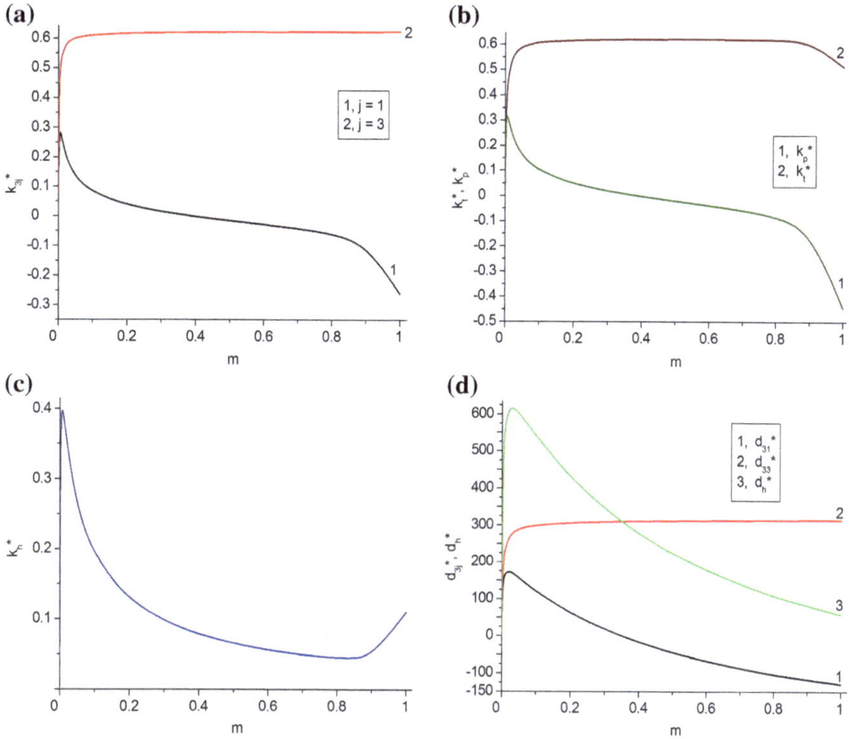

Fig. 2.3 Volume-fraction dependences of ECFs k_{3j}^* (**a**), k_t^*, k_p^* (**b**) and k_h^* (**c**), and piezoelectric coefficients d_{3j}^* and d_h^* (in pC/N) of the 1–3-type PZT-5H FC/auxetic polyethylene composite. Schematic of the composite is shown in Fig. 2.2

the non-monotonic behaviour of ECFs k_{31}^*, k_p^* and k_h^*. Large values of k_h^* are a result of the soft auxetic polymer matrix and relatively small dielectric permittivity $\varepsilon_{33}^{*\sigma}$ at $m \ll 1$.

Thus, the electromechanical coupling of the studied FC/polymer composites is highly dependent on the elastic properties of the components and on the piezoelectric activity of FC, and this dependence is important in a wide volume-fraction range ($0.01 \leq m \leq 0.99$). The above-considered composites are of interest due to the large anisotropy of ECFs, considerable electromechanical coupling and the relatively large hydrostatic ECF.

2.5.2 *Single Crystal/Polymer*

The 1–3 relaxor-ferroelectric SC/polymer composite (Fig. 2.4) contains a system of SC rods in the form of rectangular parallelepipeds which are continuous in the OX_3

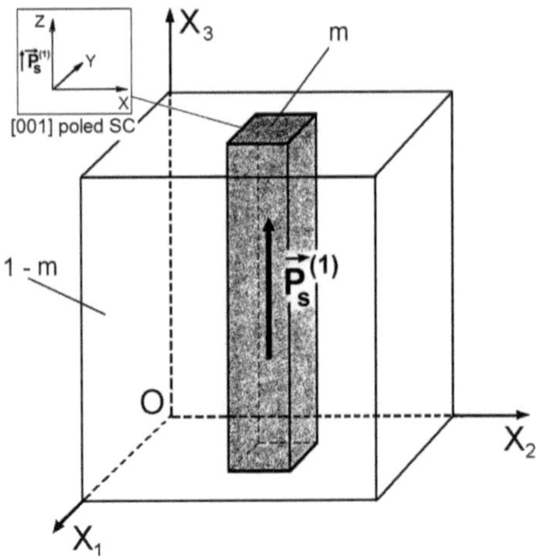

Fig. 2.4 Schematic of the 1–3 SC/polymer composite. $(X_1X_2X_3)$ is the rectangular co-ordinate system. m and $1 - m$ are volume fractions of SC and polymer, respectively, x, y and z are main crystallographic axes of SC

direction, having a square base and characterised by a square arrangement in the (X_1OX_2) plane. It is assumed that the rods are regularly distributed in a continuous polymer matrix and poled along the [001] direction in the cubic (perovskite) unit cell. In the [001]-poled PMN–*x*PT and PZN–*x*PT SCs, the main crystallographic axes x, y and z are parallel to the following perovskite unit-cell directions: x || [001], y || [010] and z || [001]. Each SC rod is characterised by a spontaneous polarisation vector $\boldsymbol{P}_s^{(1)} \uparrow\uparrow OX_3$. The prediction of the effective electromechanical properties of the studied 1–3 composite (Fig. 2.4) is carried out within the framework of the matrix approach [6, 18]. The full sets of electromechanical constants of the SC and polymer components are taken from Tables 1.1 and 2.7, respectively.

Data from Tables 2.10, 2.11 and Fig. 2.5 show that the polymer component influences the volume-fraction behaviour of all of the studied ECFs, and the auxetic polyethylene promotes very large values of hydrostatic parameters. The larger values of maxk_t^* are related to the composite based on the PMN–0.33PT SC despite relatively small differences between the piezoelectric coefficients $d_{3j}^{(1)}$ of the PMN–0.33PT and PZN–0.07PT SCs. This feature may be the result of the role of elastic and dielectric properties of the SC component in the formation of the thickness ECF.

In the case of the 1–3-type FC/auxetic polymer composite (see Sect. 2.5.1), the piezoelectric coefficient d_{31}^* (curve 1 in Fig. 2.5d) influences the volume-fraction behaviour of the ECFs k_{31}^*, k_p^* and k_h^* (Fig. 2.5a–c). Large positive values of d_{31}^* lead to a very large hydrostatic piezoelectric coefficient d_h^* (curve 3 in Fig. 2.5d) that leads to large values of k_h^*, especially at $m < 0.2$, i.e., at relatively small dielectric permittivity $\varepsilon_{33}^{*\sigma}$ of the composite (see also (2.8)).

Of specific interest are novel lead-free 1–3-type SC/auxetic polymer composites [19]. As is known, in the last decade environmental concerns have led to attempts to eliminate lead-based materials from various consumer items. Undoubtedly, this performance is important for piezoelectric energy-harvesting and hydrophone

Table 2.10 Features of the volume-fraction dependence of ECFs and anisotropy factors of 1–3 composites based on PMN–0.33PT SC

k_{31}^*	k_{33}^*	ζ_{k3j}^*	k_t^*	k_p^*	$\zeta_{t\text{-}p}^*$	k_h^*
PMN–0.33PT SC/araldite composite						
–[a]	–[b]	–[b]	maxk_t^* = 0.861 (m = 0.559)	–[a]	min$\zeta_{t\text{-}p}^*$ = –1.38 (m = 0.151)	maxk_h^* = 0.223 (m = 0.167)
PMN–0.33PT SC/polyurethane composite						
–[a]	–[b]	minζ_{k3j}^* = –2.71 (m = 0.026)	maxk_t^* = 0.897 (m = 0.544)	–[a]	min$\zeta_{t\text{-}p}^*$ = –1.49 (m = 0.212)	maxk_h^* = 0.214 (m = 0.095)
PMN–0.33PT SC/elastomer composite						
–[a]	–[b]	minζ_{k3j}^* = –3.29 (m = 0.291)	maxk_t^* = 0.939 (m = 0.535)	mink_p^* = –0.667 (m = 0.025), maxk_p^* = –0.485 (m = 0.405)	min$\zeta_{t\text{-}p}^*$ = –1.93 (m = 0.414)	maxk_h^* = 0.130 (m = 0.017)
PMN–0.33PT SC/auxetic polyethylene composite						
maxk_{31}^* = 0.537 (m = 0.011), k_{31}^* = 0 (m = 0.351)	–[b]	]–∞; ∞[	maxk_t^* = 0.951 (m = 0.524)	k_p^* = 0 (m = 0.351)	]–∞; ∞[	maxk_h^* = 0.738 (m = 0.020), mink_h^* = 0.111 (m = 0.951)

[a]Monotonic decrease
[b]Monotonic increase

applications, piezoelectric transducers, sensors and actuators. The need for lead-free ferroelectrics gives rise to the challenge of finding an alternative, to conventional lead-based FCs and ferroelectric SCs, which exhibit high piezoelectric activity playing the key role in piezo-active composites. Recent experimental studies of perovskite-type ferroelectric solid solutions such as $(K_{0.562}Na_{0.438})(Nb_{0.768}Ta_{0.232})O_3$ (KNN-T) [20] and $[Li_x(K_{0.501}Na_{0.499})_{1-x}](Nb_{0.660}Ta_{0.340})O_3$ (KNN-TL) [21] suggest that these lead-free materials in the form of domain-engineered SCs can be used in composites [19] due to their relatively large piezoelectric coefficients.

It is assumed that the composite sample contains a system of ferroelectric SC rods in the form of rectangular parallelepipeds (Fig. 2.4) which are surrounded by an auxetic polymer medium. The SC rods are continuous in the OX_3 direction, having a square base and characterised by a square arrangement in the (X_1OX_2) plane. The rods are poled along the [001] direction of the perovskite unit cell, and the main crystallographic axes x, y and z of each SC rod obey the conditions x || [001], y || [010] and z || [001]. The polymer component is an isotropic auxetic microporous polyethylene (PE-n, n = 1, 2,..., 9) with a negative Poisson's ratio [22]. It is assumed that the relative dielectric permittivity of PE-n equals 2.3, as is the case for monolithic PE [23]. Because of the microporous structure of the

Table 2.11 Features of the volume-fraction dependence of ECFs and anisotropy factors of 1–3 FC/polymer composites based on PZN–0.07PT SC

k_{31}^*	k_{33}^*	ζ_{k3j}^*	k_t^*	k_p^*	$\zeta_{t\text{-}p}^*$	k_h^*
PZN–0.07PT SC/araldite composite						
–[a]	–[b]	–[b]	$\max k_t^* =$ 0.765 (m = 0.522)	–[a]	$\min \zeta_{t\text{-}p}^* =$ –1.27 (m = 0.121)	$\max k_h^* =$ 0.193 (m = 0.179)
PZN–0.07PT SC/polyurethane composite						
–[a]	–[b]	–[b]	$\max k_t^* =$ 0.823 (m = 0.515)	–[a]	$\min \zeta_{t\text{-}p}^* =$ –1.38 (m = 0.190)	$\max k_h^* =$ 0.190 (m = 0.112)
PZN–0.07PT SC/elastomer composite						
$\min k_{31}^* =$ –0.335 (m = 0.033), $\max k_{31}^* =$ –0.299 (m = 0.245)	–[b]	$\min \zeta_{k3j}^* =$ –3.09 (m = 0.258)	$\max k_t^* =$ 0.897 (m = 0.513)	$\min k_p^* =$ –0.629 (m = 0.033), $\max k_p^* =$ –0.506 (m = 0.368)	$\min \zeta_{t\text{-}p}^* =$ –1.77 (m = 0.385)	$\max k_h^* =$ 0.119 (m = 0.020)
PZN–0.07PT SC/auxetic polyethylene composite						
$\max k_{31}^* =$ 0.512 (m = 0.014), $k_{31}^* = 0$ (m = 0.335)	–[b]	]–∞; ∞[	$\max k_t^* =$ 0.920 (m = 0.508)	$k_p^* = 0$ (m = 0.335)	]–∞; ∞[	$\max k_h^* =$ 0.684 (m = 0.026), $k_h^* = 0$ (m = 0.991)

[a]Monotonic decrease
[b]Monotonic increase

PE-n [22] matrix, the composite is described by a 1–3–γ connectivity, i.e., it can be termed a '1–3-type composite'.

The effective electromechanical properties of the ferroelectric SC/auxetic PE-n composite with planar interfaces are determined using the matrix method [18] and full sets of electromechanical constants of SCs (Table 2.12) and auxetic PE-n [22]. It is stated that in the KNN-T SC/auxetic PE-n composite, largest $\max k_t^* = 0.830$ (at m = 0.595) is achieved in the presence of the PE-5 component, and largest $\max k_h^* = 0.477$ (at m = 0.095) is related to the case of the PE-9 matrix. In the KNN-TL SC/auxetic PE-n composite, largest $\max k_t^* = 0.785$ (at m = 0.543) is achieved in the presence of the PE-5 component, and largest $\max k_h^* = 0.435$ (at m = 0.092) is related to the case of the PE-8 matrix [19]. It is seen that PE-5 shows the highest 'activity' and exhibits the largest ECF values irrespective of the type of SC. Among PE-n, isotropic PE-5 is characterised by the largest elastic compliance s_{11} [22] which strongly influences the piezoelectric response along the OX_3 direction. PE-8 and PE-9 have Poission's ratios $\nu^{(2)} < -0.5$, however the s_{11} of PE-8 is larger and leads to a stronger piezoelectric response along the OX_3 axis (Fig. 2.4). It should be added that the 1–3-type composite with the aforementioned SC volume fractions ($m > 0.09$) can be manufactured by methods [17, 18] such as rod placement, injection moulding, robocatsing, etc.

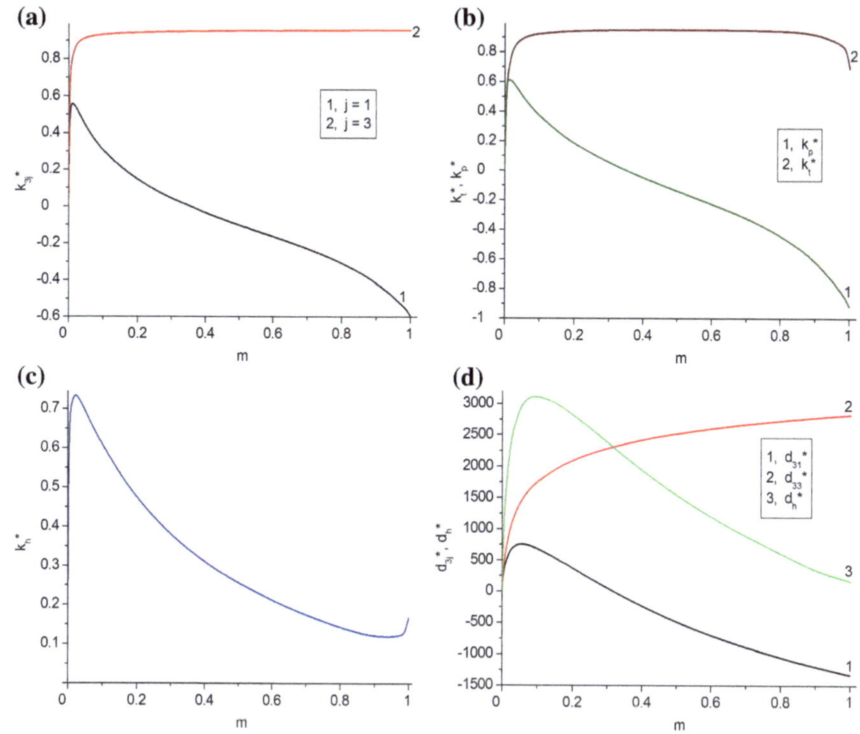

Fig. 2.5 Volume-fraction dependences of ECFs k_{3j}^* (**a**), k_t^*, k_p^* (**b**) and k_h^* (**c**), and piezoelectric coefficients d_{3j}^* and d_h^* (in pC/N) of the 1–3 PMN–0.33PT SC/auxetic polyethylene composite. Schematic of the composite is shown in Fig. 2.4

Table 2.12 Experimental values of elastic compliances s_{pq}^E (in 10^{-12} Pa^{-1}), piezoelectric coefficients d_{ij} (in pC/N) and dielectric permittivities ε_{kk}^σ of domain-engineered lead-free SCs (4*mm* symmetry) at room temperature

SC	s_{11}^E	s_{12}^E	s_{13}^E	s_{33}^E	s_{44}^E	s_{66}^E	d_{31}	d_{33}	d_{15}	$\frac{\varepsilon_{11}^\sigma}{\varepsilon_0}$	$\frac{\varepsilon_{33}^\sigma}{\varepsilon_0}$
KNN-T [20]	11.9	–4.30	–5.60	15.5	12.0	10.7	–77.0	162	45.0	291	267
KNN-TL [21]	17.2	–5.4	–10.7	27.0	15.4	13.9	–163	354	171	1100	790

The use of auxetic polymers leads to a non-monotonic behaviour of the piezoelectric coefficient d_{31}^*, and the equality

$$d_{31}^*(m^*) = 0 \tag{2.9}$$

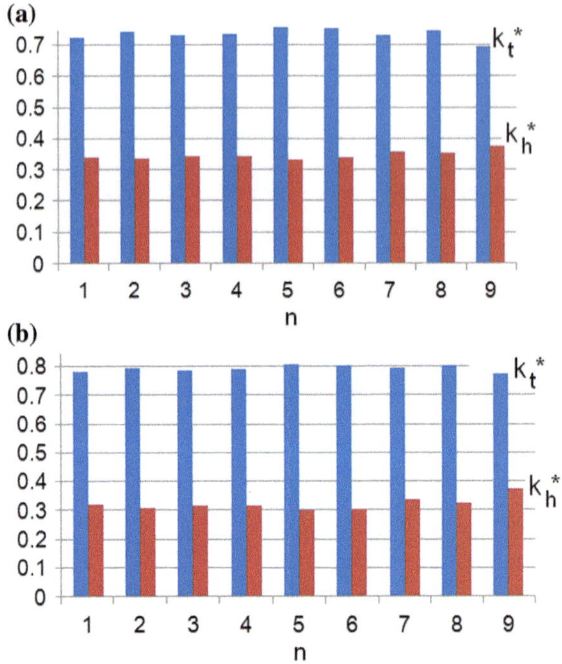

Fig. 2.6 ECFs k_t^* and k_h^* which are related to the 1–3-type KNN-T SC/auxetic PE-*n* composite at $m = m^*$ (**a**) and to the 1–3-type KNN-TL SC/auxetic PE-*n* composite at $m = m^*$, where m^* is the volume fraction of SC from (2.9) (reprinted from paper by Topolov and Bowen [19], with permission from Elsevier)

leads to the piezoelectric anisotropy $d_{33}^*/|d_{31}^*| \to \infty$ and the transverse ECF $k_{31}^* = 0$. At the volume fraction $m = m^*$, the hydrostatic piezoelectric coefficients would be $d_h^* = d_{33}^*$ and $g_h^* = g_{33}^*$, and the hydrostatic ECF would be written as $k_h^* = d_{33}^*\left(s_h^{*E}\varepsilon_{33}^{*\sigma}\right)^{-1/2}$. ECFs k_h^* and k_t^* undergo minor changes (Fig. 2.6) when replacing PE-*n* at $m = m^*$ from condition (2.9). Due to the 1–3-type composite structure (Fig. 2.4), the condition $k_t^* = k_{33}^*$ is valid at $m = m^*$ with an accuracy to 3 %.

The studied 1–3 SC/auxetic polymer composites are of value due to the strong electromechanical coupling, due to the large anisotropy of ECFs and due to the considerable hydrostatic piezoelectric response. The validity of the condition (2.9) opens up possibilities to manufacture novel lead-free composites with the infinitely large piezoelectric anisotropy and large ECFs.

2.5.3 Single Crystal/Ceramic

The ferroelectric SC/FC composites based on $PbTiO_3$ have been studied [24] by experimental methods. Some features of the effective electromechanical properties of these piezo-active materials were studied [25, 26] taking into account the composition, domain structure in the components and microstructure. Samples of the 0–3 $Pb(Ni_{1/3}Nb_{2/3})O_3 - PbTiO_3$ SC/$Pb(Ni_{1/3}Nb_{2/3})O_3 - Pb(Zr, Ti)O_3$ FC composite were manufactured by advanced hybrid sintering that combines microwave heating and hot pressing [27] and show good characteristics in the presence of the relaxor-ferroelectric component (namely, isolated inclusions of $Pb(Ni_{1/3}Nb_{2/3})O_3 - PbTiO_3$ SC). This composite material is characterised by a good interface between the SC and FC components so that neither cracks nor re-solidification of melted material are observed at the boundaries separating these components [27].

The modified $PbTiO_3$ FCs are used as matrix components in the 0–3 SC/FC composites. These FCs are characterised [28, 29] by a moderate piezoelectric activity (*e.g.*, the piezoelectric coefficient d_{33} can be about 50 times less than that of the domain-engineered PMN–*x*PT or PZN–*y*PT SCs, see Tables 1.1 and 1.2) and a large anisotropy of the piezoelectric coefficients (i.e., $d_{33}/|d_{31}| \geq 10$) or ECFs (i.e., $k_{33}/|k_{31}| \geq 10$ or $k_t/|k_p| \geq 10$). The presence of the highly anisotropic piezoelectric matrix opens up new possibility of the variation of the electromechanical coupling, piezoelectric effect and related parameters.

An alternative to the relaxor-ferroelectric SCs could be textured FCs [30] wherein a high degree of crystallographic texture improves the piezoelectric properties of the heterogeneous materials as a whole. According to experimental data from work [30], textured PMN–*x*PT FCs with compositions near the morphotropic phase boundary ($0.25 < x < 0.35$) are processed by the homo-template grain growth, and a favourable orientation of templates is achieved during the tape-casting process. An important feature of such FC samples is the small influence of connectivity and spatial distribution of the phases on the effective electromechanical properties.

Our further consideration of the SC/FC composite is carried out within the framework of the model of a 0–3 composite with spheroidal inclusions (Fig. 2.7). The shape of each inclusion is characterised by the equation

$$(x_1/a_1)^2 + (x_2/a_1)^2 + (x_3/a_3)^2 = 1 \qquad (2.10)$$

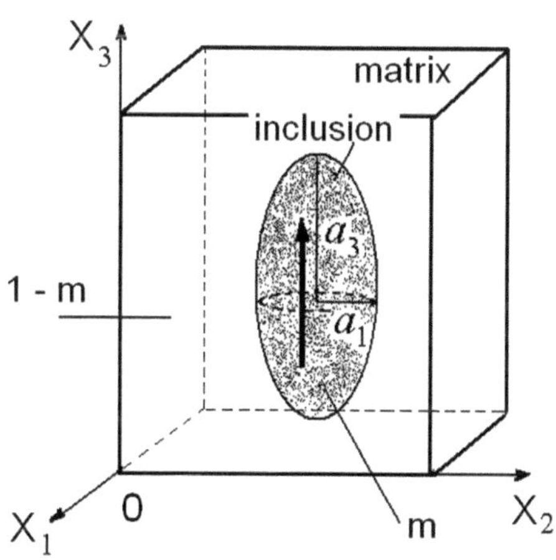

Fig. 2.7 Schematic of the 0–3 SC/FC composite. $(X_1X_2X_3)$ is the rectangular co-ordinate system. *m* and $1 - m$ are volume fractions of SC (inclusion) and FC (matrix), respectively. a_1 and a_3 are semi-axes of the spheroidal inclusion, and the *arrow* therein indicates the direction of the spontaneous polarisation vector in the SC component

in the rectangular co-ordinate system ($X_1X_2X_3$). The SC inclusions are uniformly aligned in the FC matrix and form a periodic structure. The main crystallographic axes x, y and z of each SC inclusion are parallel to the co-ordinate axes OX_1, OX_2 and OX_3, respectively, and the poling axis is z || [001] || OX_3. In accordance with (2.10), the semi-axes of each inclusion are a_1, $a_2 = a_1$ and a_3. The remanent polarisation of the FC matrix is also parallel to the OX_3 axis. Taking into account the electromechanical interaction between the piezo-active components in the composite, we determine its effective electromechanical properties $\Pi^* = \Pi^*(m, \rho)$ by means of the effective field method [6, 18] where $\rho = a_1/a_3$ is the aspect ratio of the inclusion. In a limiting case of $\rho = 0$, we have a system of infinitely long SC rods parallel to the poling axis OX_3 and 1–3 connectivity of the composite.

Below we consider two examples of ECFs and their anisotropy in the SC/FC composites. In the first example, the PZ 34 FC matrix is characterised by a large anisotropy of the piezoelectric coefficients (see Table 1.2), i.e., $d_{33}/|d_{31}| \approx 15$, and the piezoelectric coefficient d_{33} is much lower than d_{33} of the SC component. This difference and the elastic anisotropy of both the components leads to the considerbale anisotropy of ECFs. For example, conditions

$$\left|\zeta^*_{k3j}\right| \geq 5 \tag{2.11}$$

and

$$\left|\zeta^*_{kt-p}\right| \geq 5 \tag{2.12}$$

are valid in relatively wide ranges at $\rho < 0.2$ (Fig. 2.8).[1] Such a feature means that the addition of fairly oblate SC inclusions with high piezoelectric activity promote a strong electromechanical coupling along the poling axis OX_3. Differences between graphs in Fig. 2.8a, b are mainly accounted for by the elastic anisotropy of the SC component (cf. data on PMN–0.33PT and PZN–0.07PT SC from Table 1.1). A wider range of volume fractions m (Fig. 2.7) is related to inequality (2.12). We remind the reader that the anisotropy factor ζ^*_{kt-p} from (2.7) strongly depends on the elastic properties of the composite, and these properties are sensitive to changes in both m and ρ.

If we replace the PZ 34 FC matrix with a PZT-type matrix, significant changes in the electromechanical interaction between the components are observed. The piezoelectric anisotropy of the PZT-type matrix does not result in a pronounced electromechanical coupling in the composite along its poling axis. This results in

[1] In our evaluations of ECFs and their anisotropy, we varied the volume fraction m of the SC inclusions from 0 to 0.5. The upper limit $m = \pi/6$ at the regular arrangement of spherical inclusions ($\rho = 1$ in the composite structure shown in Fig. 2.7) whose centres form a simple cubic lattice. We add that in the case of $\rho = 0$ (1–3 composite with cylindrical rods), the upper limit is $m = \pi/4$.

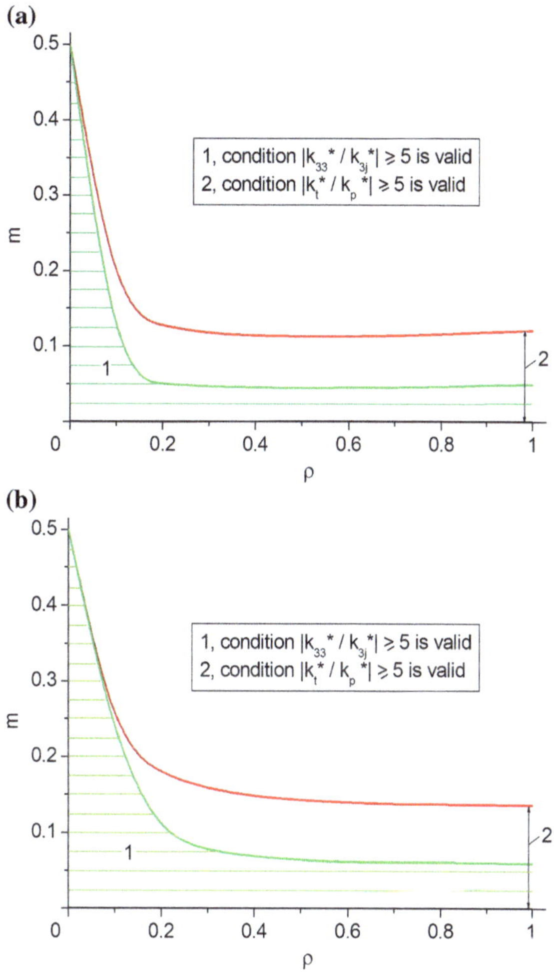

Fig. 2.8 Regions of aspect ratio ρ and volume fraction m for which conditions (2.11) and (2.12) (regions 1 and 2, respectively) are valid in the PMN–0.33PT SC/PZ34 FC composite (**a**) and PZN–0.07PT SC/PZ34 FC composite (**b**). Schematic of the 0–3 composite at $0 < \rho < 1$ is shown in Fig. 2.7

smaller values of $\left|\zeta_{k3j}^{*}\right|$ and $\left|\zeta_{kt-p}^{*}\right|$ in wide ranges of m and ρ, so that inequalities (2.11) and (2.12) do not hold for the SC/PZT-type FC composites.

The second example is concerned with ECFs of the 1–3 PMN–0.33PT SC/FC composite with circular rods (Fig. 2.9). If we vary the electromechanical properties of the FC component, its piezoelectric activity remains significantly lower than that of the SC component (cf. data from Tables 1.1 and 1.2). Changes in the anisotropy of the piezoelectric coefficients d_{ij} and elastic compliances s_{ab}^{E} of the FC give rise to changes in the mutual arrangement of curves at $m < 0.3$ (see Fig. 2.9), and in this range the FC matrix plays a dominant role in the anisotropy of the ECFs. At 0. $3 < m < 0.5$ we see minor changes in the curves, and therefore, the influence of the FC matrix becomes restricted. At $0.5 < m < 0.7$ the SC rods play a key role in the formation of the effective electromechanical properties in the composite, and

Fig. 2.9 Volume-fraction behaviour of ECFs of 1–3 SC/FC composites with cylindrical rods ($\rho = 0$) as follows: PMN–0.33PT SC/PZT-5H FC (**a**), PMN–0.33PT/modified $PbTiO_3$ (I) FC (**b**) and PMN–0.33PT SC/PMN–0.35PT FC (**c**)

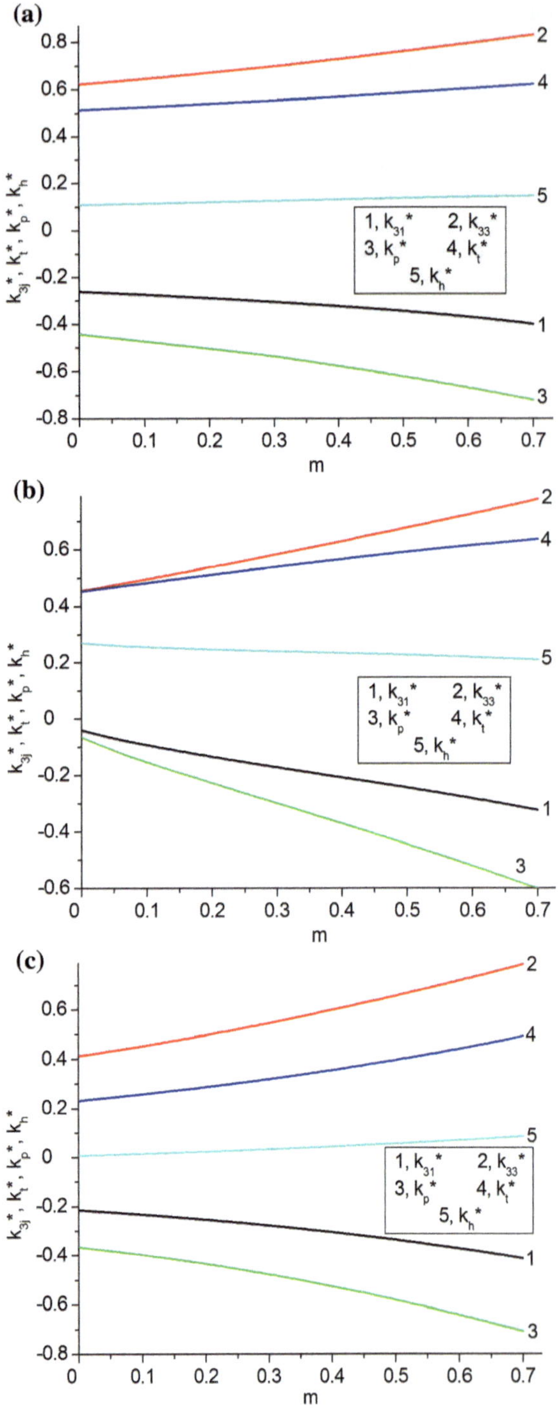

differences between the mutual arrangement of curves in Fig. 2.9 at various FC components become minor. It should be added that conditions (2.11) and (2.12) are valid in the presence of the highly anisotropic modified $PbTiO_3$ FC (Fig. 2.9b) at $m < 0.15$, and an increase of the volume fraction m leads to weakening of the anisotropy of electromechanical coupling due to the SC component.

A specific feature of graphs in Fig. 2.9 is that it consists of a weak volume-fraction dependence of the hydrostatic ECF k_h^* (see curve 5). This ECF is determined in accordance with (2.8), and a weak dependence is observed due to the combination of the hydrostatic piezoelectric coefficient d_h^* and elastic compliance s_h^{*E} at increasing dielectric permittivity $\varepsilon_{33}^{*\sigma}$. In our opinion, any changes in d_h^* are almost compensated for by changes in $\left(s_h^{*E}\varepsilon_{33}^{*\sigma}\right)$, and the last product varies in a relatively narrow range due to relatively small differences between both the elastic and dielectric properties of the SC and FC components.

Thus, the results discussed in Sect. 2.5.3 show that the role of the elastic and piezoelectric properties of both the SC and FC components is to be taken into consideration at the prediction of the effective properties and ECFs of the studied 0–3 and 1–3 composite. The possibility of a large anisotropy of ECFs (see conditions (2.11) and (2.12)) is of value for piezoelectric transducers exploiting the longitudinal oscillation mode and promoting effective energy conversion and harvesting only in the poling direction.

2.6 Porous Composite Materials

In Sect. 2.6 we consider ECFs and related parameters of porous piezo-active composites based on ferroelectrics. A system of air pores in a piezo-passive polymer matrix influences the elastic and dielectric properties of a medium that surrounds the ferroelectric inclusion (particle, rod or layer). The formation of a regular porous structure in this polymer matrix leads to an anisotropy of its elastic and dielectric properties [6, 31–33] that influence the electromechanical coupling in the composite and the anisotropy of its ECFs. Below we analyse examples of the large anisotropy of ECFs in three-component 1–3-type composites that represent a system of piezoelectric rods (either FC or relaxor-ferroelectric SC) in a porous polymer matrix. It is clear that the models of the 1–3 FC/polymer and SC/polymer composites (see Sects. 2.5.1 and 2.5.2) are to be modified in order to predict the effective electromechanical properties of the related 1–3-type porous composites.

2.6.1 Porous Composites Based on Ferroelectric Ceramics

The 1–3-type composite that contains a system of the FC rods in a porous polymer matrix (Fig. 2.10) is of interest due to the presence of an anisotropic piezoelectric

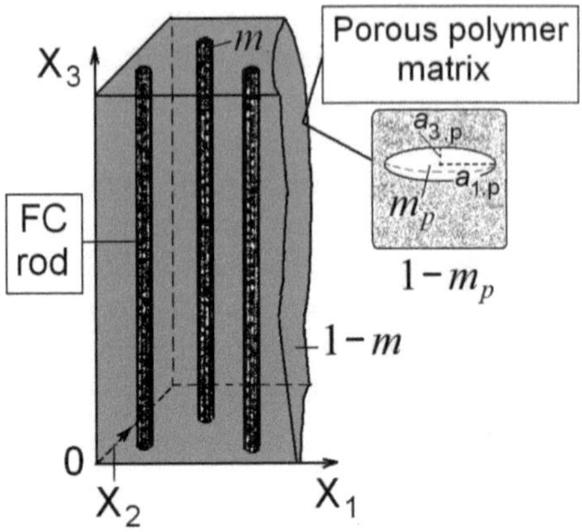

Fig. 2.10 Schematic of the 1–3–0 FC/porous polymer composite. ($X_1X_2X_3$) is the rectangular co-ordinate system, m is the volume fraction of FC, $1 - m$ is the volume fraction of porous polymer, m_p is porosity of the polymer matrix, $a_{1,p}$ and $a_{3,p}$ are semi-axes of the spheroidal pore shown in *inset* (reprinted from paper by Topolov et al. [33], with permission from Taylor & Francis)

component surrounded by an anisotropic porous medium. The FC rods are in the form of a circular cylinder with a constant radius and parallel to the co-ordinate OX_3 axis which is the poling direction of the composite sample as a whole. The FC rods are characterised by a periodical square arrangement in the (X_1OX_2) plane. The remanent polarisation of each FC rod is $\boldsymbol{P}_r^{(1)} \uparrow\uparrow OX_3$, and electrodes covering the composite sample (Fig. 2.10) are parallel to the (X_1OX_2) plane.

The matrix surrounding each FC rod contains a system of spheroidal air inclusions (pores) that are described by the equation

$$\left(x_1/a_{1,p}\right)^2 + \left(x_2/a_{1,p}\right)^2 + \left(x_3/a_{3,p}\right)^2 = 1 \qquad (2.13)$$

relative to the co-ordinate axes (OX_j). In (2.13) $a_{1,p}$ and $a_{3,p}$ are semi-axes of the spheroidal pore. The porous matrix is characterised by 3–0 connectivity in terms of work [18]. The pores are regularly distributed in the polymer matrix and occupy the sites of a simple tetragonal lattice. The shape of each pore is characterised by an aspect ratio $\rho_p = a_{1,p}/a_{3,p}$ that is constant over the composite sample. The radius or the largest semi-axis of each pore (e.g., $a_{1,p}$ for the oblate pore or $a_{3,p}$ for the prolate pore) remains considerably less than the radius of the cylindrical rod. In addition, methods for manufacturing a porous polymer material with a regular arrangement of pores having the appointed shape are known, for instance, from patents [34, 35]; these materials are also termed ‘patterned porous polymers’. The FC/porous polymer composite shown in Fig. 2.10 is described by 1–3–0 connectivity, and the effective electromechanical properties of this composite are to be regarded as a function of the volume fraction of FC m, porosity of the polymer matrix m_p and aspect ratio of the pore ρ_p.

A determination of the full set of electromechanical constants of the 1–3–0 composite is carried out as follows. At the first stage the effective properties of the polymer matrix with aligned spheroidal pores (see inset in Fig. 2.10) are determined as a function of the volume fraction of m_p and ρ_p within the framework of Eshelby’s concepts of spheroidal inclusions in heterogeneous solids [6, 18]. At the second stage we use the effective-field method [6, 18] to determine the effective electromechanical properties of the 1–3–0 composite $\Pi^* = \Pi^*(m, m_p, \rho_p)$. Following the effective-field method, we describe an electromechanical interaction in the 'FC rods—porous polymer matrix' system using a local field that acts on each piezoelectric rod. As in Sect. 2.5, we take experimental electromechanical constants of FC and polymer components from Tables 1.2 and 2.7 for further calculations.

Table 2.13 contains data on the composites based on PZT-type FCs. As is known from Tables 1.2 and 2.5, these FCs do not show a remarkable anisotropy of the piezoelectric coefficients d_{3j} and ECFs k_{3j}. The system of poled PZT-type FC rods in the porous polymer matrix promotes a pronounced piezoelectric anisotropy that also leads to a large anisotropy of the ECFs. Data from Table 2.13 suggest, that the volume-fraction ranges, wherein conditions (2.11) and (2.12) for the large anisotropy of ECFs hold, depend on porosity m_p to a considerable extent. At $m_p = 0.3$ the aforementioned range related to condition (2.11) concerning k_{3j}^* becomes very wide irrespective of the FC component. This is due to the active role of the anisotropic porous matrix (see inset in Fig. 2.10) in the formation of the electromechanical coupling in the composite. Its large elastic compliance s_{33}^{*E} is achieved in the poling direction OX_3, and the value of s_{33}^{*E} increases with increasing m_p. In contrast to ECFs k_{3j}^*, the anisotropy of k_t^* and k_p^* is achieved in narrower ranges $[m_1; m_2]$ (see the 4th column of Table 2.13) because of the strong influence of the anisotropy of elastic properties of the composite as a whole.

The elastic anisotropy of the porous polymer matrix at $\rho_p = 0.1$–10 is insufficient to promote a large anisotropy of ECFs k_t^* and k_p^*. For example, at $\rho_p = 0.1$ and $m_p = 0.1$, ratios of elastic moduli of the porous polyurethane matrix $c_{33}^{(3)}/c_{13}^{(3)} = 1.70$ and $c_{11}^{(3)}/c_{13}^{(3)} = 1.75$ are comparable to the similar ratios $c_{33}^{(1),E}/c_{13}^{(1),E}$ and $c_{11}^{(1),E}/c_{13}^{(1),E}$ of the PCR-7M FC. If the polyurethane matrix comprises the oblate pores ($\rho_p > 1$), then the elastic anisotropy of the matrix becomes more pronounced: for instance, $c_{33}^{(3)}/c_{13}^{(3)} = 1.38$ and $c_{11}^{(3)}/c_{13}^{(3)} = 4.25$ at $\rho_p = 10$ and $m_p = 0.3$, and $c_{33}^{(3)}/c_{13}^{(3)} = 1.36$ and $c_{11}^{(3)}/c_{13}^{(3)} = 29.9$ at $\rho_p = 100$ and $m_p = 0.3$. The large difference between $c_{11}^{(3)}/c_{13}^{(3)}$ and $c_{33}^{(3)}/c_{13}^{(3)}$ gives rise to a considerable re-distribution of internal fields in the composite sample and leads to the large anisotropy of the piezoelectric coefficients, ECFs and related parameters [33].

Table 2.14 summarises results on a correlation between minima of the anisotropy factors of the composite and ratios of elastic moduli of its porous matrix. It is seen that a competition between the elastic moduli related to the poling direction OX_3 and remaining directions OX_j (j = 1 and 2) influences the anisotropy of ECFs. We add that the studied 1–3–0 composite is characterised by a relatively high piezoelectric activity in a wide range of volume fractions of FC m irrespective of the aspect ratio ρ_p and volume fraction m_p of air inclusions, i.e., the piezoelectric coefficient $d_{33}^* \sim 100\,\text{pC/N}$ due to the presence of the PZT-type FC rods. This performance of the 1–3–0 composite implies its advantage over the conventional

Table 2.13 Volume-fraction ranges $[m_1; m_2]$ wherein conditions (2.11) and (2.12) for the large anisotropy of ECFs in 1–3–0 FC/porous polymer composites are valid

ρ_p	m_p	Volume-fraction range $[m_1; m_2]$ related to condition (2.11)	Volume-fraction range $[m_1; m_2]$ related to condition (2.12)
ZTS-19 FC/porous polyurethane composite			
0.1	0.1	[0.124; 0.700]	–
	0.2	[0.080; 0.700]	–
	0.3	[0.054; 0.700]	–
1	0.1	[0.137; 0.700]	–
	0.2	[0.097; 0.700]	–
	0.3	[0.069; 0.700]	–
10	0.1	[0.050; 0.700]	–
	0.2	[0.021; 0.700]	–
	0.3	[0.011; 0.700]	[0.149; 0.352]
100	0.1	[0.001; 0.700]	[0.008; 0.472]
	0.2	[0.001; 0.700]	[0.001; 0.541]
	0.3	[0.001; 0.700]	[0.001; 0.588]
PZT-5H FC/porous polyurethane composite			
0.1	0.1	[0.164; 0.665]	–
	0.2	[0.101; 0.700]	–
	0.3	[0.066; 0.700]	–
1	0.1	[0.188; 0.621]	–
	0.2	[0.124; 0.700]	–
	0.3	[0.086; 0.700]	–
10	0.1	[0.062; 0.691]	–
	0.2	[0.025; 0.700]	–
	0.3	[0.013; 0.700]	[0.001; 0.463]
100	0.1	[0.001; 0.700]	[0.008; 0.384]
	0.2	[0.001; 0.700]	[0.001; 0.460]
	0.3	[0.001; 0.700]	[0.001; 0.542]
PCR-7M FC/porous polyurethane composite			
0.1	0.1	[0.167; 0.637]	–
	0.2	[0.101; 0.700]	–
	0.3	[0.065; 0.700]	–
1	0.1	[0.191; 0.590]	–
	0.2	[0.124; 0.692]	–
	0.3	[0.085; 0.700]	–
10	0.1	[0.063; 0.669]	–
	0.2	[0.025; 0.700]	–
	0.3	[0.013; 0.700]	–
100	0.1	[0.001; 0.700]	[0.009; 0.354]
	0.2	[0.001; 0.700]	[0.001; 0.431]
	0.3	[0.001; 0.700]	[0.001; 0.483]

Table 2.14 Minimum values of anisotropy factors Anisotropy factor ζ^*_{k3j} and ζ^*_{kt-p} of the 1–3–0 Piezo-composite PCR-7M FC/porous polyurethane composite at m_p = 0.3 and ratios that characterise the elastic anisotropy of the porous matrix in this composite (reprinted from paper by Topolov et al. [33], with permission from Taylor & Francis)

ρ_p	$\min\zeta_{k3j}*$	$\min\zeta_{kt-p}*$	$c_{33}^{(3)}/c_{13}^{(3)}$	$c_{11}^{(3)}/c_{13}^{(3)}$	$\left\|c_{33}^{(3)}/c_{11}^{(3)}\right\|/c_{13}^{(3)}$
0.1	−7.10 (m = 0.380)[a]	−3.70 (m = 0.383)	2.67	0.691	1.98
1	−6.60 (m = 0.364)	−3.07 (m = 0.385)	1.70	1.08	0.62
10	−8.91 (m = 0.205)	−4.59 (m = 0.210)	1.38	3.09	1.71
100	−22.1 (m = 0.033)	−12.2 (m = 0.033)	1.36	29.9	28.5

[a]The volume fraction of FC m, at which minimum of the anisotropy factor is achieved, is given in parentheses (see the 2nd and 3rd columns)

$PbTiO_3$-type FC with a large anisotropy of $d_{3j}^{(1)}$ and $d_{33}^{(1)} \approx 50\,\text{pC/N}$ (see electromechanical constants in Table 1.2).

2.6.2 Porous Composites Based on Relaxor-Ferroelectric Single Crystals

By modifying the 1–3 composite structure shown in Fig. 2.4, we consider a composite that contains a system of single-domain relaxor- ferroelectric SC rods in a porous polymer matrix. The SC rods are in the form of rectangular parallelepipeds and continuous in the OX_3 direction (see Fig. 2.4) which is the poling direction of the composite sample as a whole. The SC rods have a square base and are characterised by a periodical square arrangement in the (X_1OX_2) plane. The crystallographic axes x, y and z of each SC rod are parallel to the following co ordinate axes: x $\|$ OX_1, y $\|$ OX_2 and z $\|$ OX_3. Each SC rod is characterised by a spontaneous polarisation vector $\boldsymbol{P}_s^{(1)} \uparrow\uparrow OX_3$. It is assumed that the single-domain state in the SC rods is stabilised by the bias field $\boldsymbol{E} \uparrow\uparrow OX_3$. The polymer matrix of the studied composite contains a system of spheroidal air inclusions (pores) that are described by (2.13) relative to the axes of the rectangular co-ordinate system $(X_1X_2X_3)$. The pores are regularly distributed in the polymer matrix and occupy the sites of a simple tetragonal lattice. The shape of each pore is characterised by an aspect ratio $\rho_p = a_{1,p}/a_{3,p}$ and is assumed to be fixed over the volume of the composite sample. The radius or the largest semi-axis of each pore remains much less than the length of the side of the square which is the intersection of the rod by the (X_1OX_2) plane.

The above-described SC/porous polymer composite is characterised by 1–3–0 connectivity. As in the case for the manufactured 1–3 SC/polymer composites [36–39], the poling procedure is carried out after inserting aligned SC rods into the

Table 2.15 Room-temperature experimental values of elastic compliances s_{ab}^{E} (in 10^{-12} Pa^{-1}), piezoelectric coefficients d_{ij} (in pC/N) and dielectric permittivities $\varepsilon_{pp}^{\sigma}$ of single-domain PMN–0.42PT SC [40] (4*mm* symmetry)

s_{11}^{E}	s_{12}^{E}	s_{13}^{E}	s_{33}^{E}	s_{44}^{E}	s_{66}^{E}	d_{31}	d_{33}	d_{15}	$\varepsilon_{11}^{\sigma}/\varepsilon_0$	$\varepsilon_{33}^{\sigma}/\varepsilon_0$
9.43	–1.68	–6.13	19.21	35.09	12.5	–91	260	131	8627	660

matrix and covering the composite with electrodes parallel to the (X_1OX_2) plane. The effective electromechanical properties of the 1–3–0 composite are evaluated in two stages. At the first stage the effective properties of the polymer matrix with the aligned spheroidal pores are determined as a function of their volume fraction m_p and aspect ratio ρ_p within the framework of Eshelby's concepts of piezoelectric spheroidal inclusions in heterogeneous solids [6, 18]. At the second stage we apply the matrix method [6, 18] to determine the effective electromechanical properties of the 1–3–0 composite $\Pi^* = \Pi^*(m, m_p, \rho_p)$. Electromechanical constants of the SC and polymer components are taken from Tables 2.15 and 2.7, respectively. We note that the single-domain PMN–0.42PT SC is characterised by a modest piezoelectric anisotropy: according to data from Table 2.15, $d_{33}^{(1)}/d_{31}^{(1)} \approx -2.9$.

The volume-fraction (m) dependence of absolute values of the anisotropy factors ζ_{k3j}^{*} and $\zeta_{kt\text{-}p}^{*}$ and the hydrostatic ECF k_h^* of the 1–3–0 composite based on the single-domain PMN–0.42PT SC is graphically represented in Figs. 2.11 and 2.12. It is seen that both the aspect ratio ρ_p and porosity m_p influence the behaviour of the anisotropy factors in a wide m range. A transition from the highly prolate pore ($\rho_p \ll 1$) to the highly oblate pore ($\rho_p \gg 1$) in the polymer matrix leads to increasing $|\zeta_{k3j}^{*}|$ and $|\zeta_{kt\text{-}p}^{*}|$ and gives rise to sharper maxima of $|\zeta_{k3j}^{*}|$ (Figs. 2.11a and 2.12a) and $|\zeta_{kt\text{-}p}^{*}|$ (Figs. 2.11b and 2.12b). Increasing the porosity m_p also favours an increase in both $|\zeta_{k3j}^{*}|$ and $|\zeta_{kt\text{-}p}^{*}|$. Changes in ρ_p and m_p imply that the elastic anisotropy of the porous matrix undergoes changes, and they become pronounced at larger values of m_p and mainly at $\rho_p > 1$.

As follows from our analysis of elastic compliances of the porous matrix [41], a small anisotropy of elastic compliances ($s_{11}^{(3)}/s_{33}^{(3)} \approx 1$) is observed at $\rho_p < 1$ and partially at $\rho_p > 1$ (as a rule, at $m_p < 0.2$ that corresponds to a low porosity level). The formation of highly oblate pores and increasing the porosity lead to the ratio $s_{11}^{(3)}/s_{33}^{(3)} \gg 1$, and finally, to the larger anisotropy of EFCs and piezoelectric coefficients. We add that the dielectric properties of the pores at various values of ρ_p and m_p vary in a relative narrow range so that the inequality $\varepsilon_{kk}^{(3)}/\varepsilon_{kk}^{(1),\sigma} \ll 1$ holds. Thus, the elastic anisotropy of the porous matrix plays an important role in achieving the large anisotropy of ECFs of the 1–3–0 composite.

The volume-fraction (m) dependence of the hydrostatic ECF k_h^* (see Figs. 2.11c and 2.12c) is of interest due to a non-monotonic behaviour. The sharp maximum of $k_h^*(m)$ at $0 < m < 0.1$ is concerned with the role of the relatively small dielectric permittivity $\varepsilon_{33}^{*\sigma}$ and fairly large piezoelectric coefficient d_{33}^{*} [18, 41]. The diffuse minimum of k_h^* is accounted for by the elastic anisotropy of the composite and

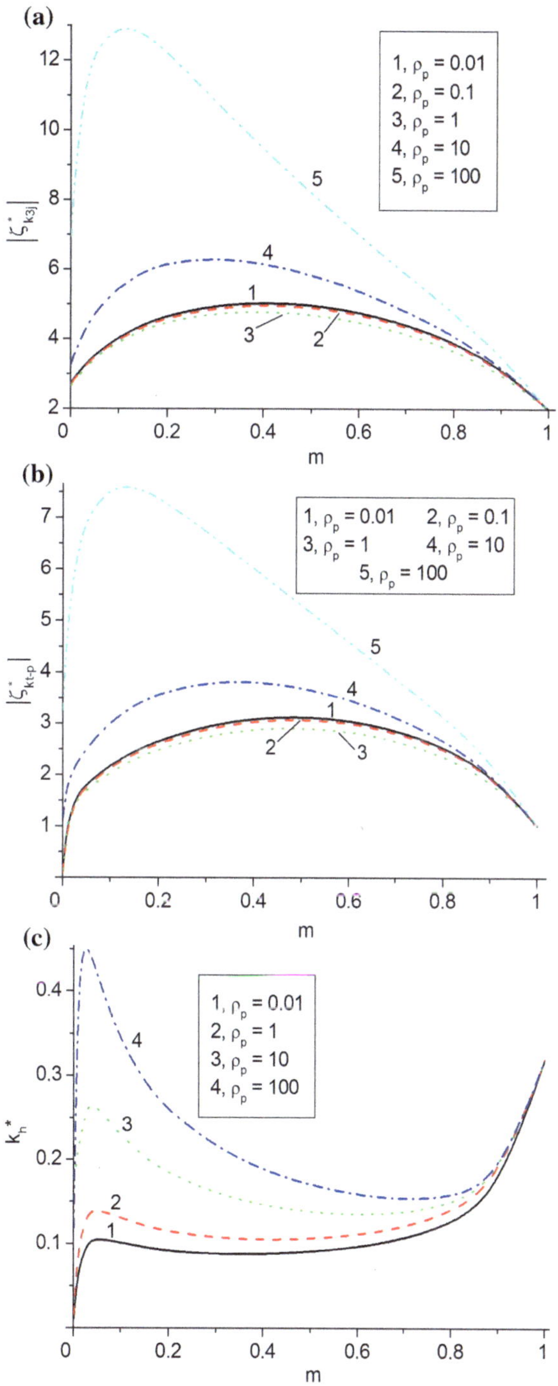

Fig. 2.11 Volume-fraction dependence of absolute values of anisotropy factors ζ^*_{k3j} (**a**) and ζ^*_{kt-p} (**b**) and hydrostatic ECF k^*_h (**c**) of the 1–3–0 single-domain PMN–0.42PT SC/porous polyurethane composite at $m_p = 0.1$ (reprinted from paper by Topolov et al. [41], with permission from Elsevier)

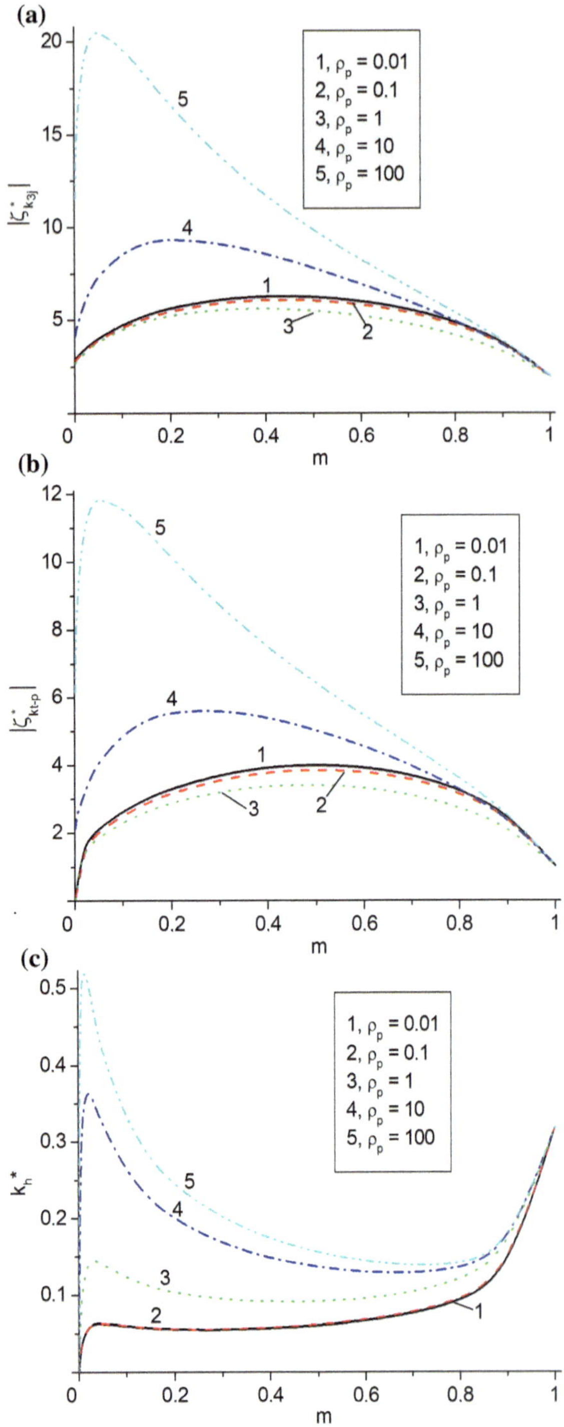

Fig. 2.12 Volume-fraction dependence of absolute values of anisotropy factors ζ^*_{k3j} (**a**) and ζ^*_{kt-p} (**b**) and hydrostatic ECF k^*_h (**c**) of the 1–3–0 single-domain PMN–0.42PT SC/porous polyurethane composite at $m_p = 0.3$ (reprinted from paper by Topolov et al. [41], with permission from Elsevier)

changes in the balance of its elastic compliances s_{ab}^{*E}. These changes become more appreciable at $\rho_p > 1$, as the elastic anisotropy of the porous matrix increases considerably. Taking into account (2.6) and (2.8), we represent the hydrostatic ECF k_h^* in terms of the anisotropy factor ζ_{d3j}^* and elastic compliances s_{ab}^{*E} as follows:

$$k_h^* = k_{33}^* \left(s_{33}^{*E}/s_h^{*E}\right)^{1/2}\left[\left(\zeta_{d3j}^* - 2\right)/\zeta_{d3j}^*\right] \quad (2.14)$$

Equation (2.14) shows that the role of the elastic compliance s_{33}^{*E} and anisotropy factor ζ_{d3j}^* in forming the hydrostatic electromechanical coupling is important. Due to large values of s_{33}^{*E} and ζ_{d3j}^* in the case of the polymer matrix with oblate pores at $\rho_p > 1$, one can achieve values of $k_h^* \approx 1.5\ k_h^{(1)}$ (see curve 4 in Fig. 2.11c and curves 4 and 5 in Fig. 2.12c). As follows from Figs. 2.11c and 2.12c, large values of k_h^* are observed at relatively small volume fractions of SC, i.e., at $m < 0.1$. We mention for comparison that Choy et al. [42] manufactured a 1–3 FC/polymer composite with low volume fractions of FC rods (approximately 0.033, 0.066 and so on), and therefore, a similar composite with volume fractions of SC $0.03 < m < 0.1$ can be manufactured.

One of important advantages of the studied 1–3–0 composite over conventional $PbTiO_3$-type FCs [28, 29] with a large anisotropy of the piezoelectric coefficients d_{3j} consists in the fact, that even at relatively small volume fractions of SC ($m \approx 0.05$), the piezoelectric coefficient of the composite d_{33}^* is about 3–4 times more [8] than d_{33} of the $PbTiO_3$-type FCs (see Table 1.2). The stronger electromechanical coupling in the studied 1–3–0 composite in comparison to the $PbTiO_3$-type FCs is achieved due to the presence of the relaxor-based SC component with higher piezoelectric activity (see data from Table 2.15). Moreover, the porous polymer component with a high elastic anisotropy due to the microgeometry of the air pores (see inset in Fig. 2.10) strongly influences the anisotropy factors and hydrostatic electromechanical coupling of the composite. Data shown in Figs. 2.11 and 2.12 and the aforementioned advantages are important for piezoelectric energy-harvesting, transducer, hydrophone, and other applications of the 1–3–0 composite based on the single domain PMN–0.42PT SC.

2.7 Conclusion

Chapter 2 has been devoted to the characterisation of electromechanical coupling in modern piezo-active composites based on either FCs or relaxor-ferroelectric SCs. The main results of the study on ECFs and their anisotropy enable us to conclude as follows.

(i) Data on the electromechanical properties and ECFs of piezoelectric and ferroelectric SCs and poled FCs (Sects. 2.1 and 2.2) show that the large anisotropy of ECFs is observed in a restricted group of FCs of the $PbTiO_3$ type. The relaxor-ferroelectric SCs with engineered domain structures are of interest due to high piezoelectric activity. However these SCs do not exhibit a large anisotropy of ECFs.

(ii) The system of the aligned FC (or SC) rods in the large polymer matrix (see Figs. 2.2 and 2.4) promotes an anisotropic piezoelectric effect in the 1–3 composite, and the elastic anisotropy of its matrix additionally influences the anisotropy factors (2.6) and (2.7) and hydrostatic ECF k_h^* from (2.8). Equations (2.6) and (2.7) show that the elastic anisotropy of the composite plays the important role in the formation of the large anisotropy of ECFs. The hydrostatic piezoelectric response and related ECF from (2.8) depend on the anisotropy of the piezoelectric coefficients d_{3j}^* and elastic compliances s_{ab}^{*E} of the composite. The aforementioned parameters of the composite strongly depend on the volume fraction of the FC (or ferroelectric SC) component, as well as on the elastic and dielectric properties of the components.

(iii) An auxetic polymer matrix strongly influences the elastic and piezoelectric anisotropy of the composite and its electromechanical coupling due to the negative Poisson's ratio of the polymer. The presence of volume fractions of the FC (or SC) component, at which $k_{31}^* = 0$ or $k_p^* = 0$, opens up new possibilities of the variation of the ECFs and their anisotropy irrespective of the electromechanical properties of the rod. The novel lead-free 1–3-type SC/auxetic PE-*n* composites can be regarded as advanced piezo-active materials with ECFs $k_h^* \approx 0.3-0.4$ and $k_t^* \approx 0.7-0.8$ at $k_{31}^* = 0$ and $d_{31}^* = 0$.

(iv) Comparing data on the related 1–3 and 0–3 SC/FC composites, we state that 1–3 connectivity provides more advantages due to the continuous distribution of the piezoelectric component in the polymer matrix along the poling direction. The FC matrix, being piezoelectric but with lower activity in comparison to that of the SC rod, influences the anisotropy of ECFs in the 1–3 SC/FC composites to a lesser degree as compared to the performance of the 1–3 SC/polymer composites. Such a restricted influence is concerned with a relatively small difference between elastic compliances of the SC and FC components. Since differences between the elastic compliances of the FC (or SC) and polymer components are large, the anisotropy of ECFs in the composite becomes considerable.

(v) The system of oblate spheroidal air pores in the matrix (see inset in Fig. 2.10) serves as an additional stimulus to increase the anisotropy of ECFs and to improve the hydrostatic piezoelectric response of the 1–3-type composite with the porous polymer matrix. Volume-fraction ranges $[m_1; m_2]$ that correspond to a validity of conditions (2.11) and (2.12) for a large anisotropy of ECFs (see Table 2.13) in the 1–3-type composite based on the PZT-type FC with the moderate anisotropy of piezoelectric and elastic properties suggest that the porous polymer matrix is mainly responsible for the anisotropy of the effective electromechanical properties in a wide volume-fraction range. Hereby ratios of elastic moduli of the porous polymer matrix (see Table 2.14) are to be taken into consideration at variations of the aspect ratio of pores ρ_p and porosity m_p.

In general, data from Chap. 2 can be taken into account to predict anisotropic electromechanical coupling and manufacture the piezo-active composites with the large anisotropy of ECFs. Knowledge of the physical and microgeometric factors that promote validity of conditions (2.11) and (2.12) would be of benefit for design of novel piezoelectric energy-harvesting composite materials with preferable directions for the conversion and propagation of energy.

References

1. Ikeda T (1990) Fundamentals of piezoelectricity. Oxford University Press, Oxford
2. Zheludev IS (1971) Physics of crystalline dielectrics. In: Electrical properties, vol. 2. Plenum, New York
3. Berlincourt DA, Cerran DR, Jaffe H (1964) Piezoelectric and piezomagnetic materials and their function in transducers. In: Mason W (ed) Physical acoustics. Principles and methods. Methods and devices, vol 1. Academic Press, New York, pp 169–270
4. Blistanov AA, Bondarenko VS, Perelomova NV, Strizhevskaya FN, Chkalova VV, Shaskol'skaya MP (1982) Acoustic crystals (Handbook). Nauka, Moscow (in Russian)
5. Xu Y (1991) Ferroelectric materials and their applications. North-Holland, Amsterdam
6. Topolov VYu, Bisegna P, Bowen CR (2014) Piezo-active composites. Orientation effects and anisotropy factors. Springer, Heidelberg
7. Fesenko EG, Gavrilyachenko VG, Semenchev AF (1990) Domain structure of multiaxial ferroelectric crystals. Rostov University Press, Rostov-on-Don (in Russian)
8. Turik AV (1970) Elastic, piezoelectric, and dielectric properties of single crystals of $BaTiO_3$ with a laminar domain structure. Sov Phys Solid State 12:688–693
9. Aleshin VI (1990) Domain-orientation contribution into constants of ferroelectric polydomain single crystal. Zh Tekh Fiz 60:179–183 (in Russian)
10. Park S-E, Shrout TR (1997) Ultrahigh strain and piezoelectric behavior in relaxor based ferroelectric single crystals. J Appl Phys 82:1804–1811
11. Dammak H, Renault A-É, Gaucher P, Thi MP, Calvarin G (2003) Origin of the giant piezoelectric properties in the [001] domain engineered relaxor single crystals. Jpn J Appl Phys 1(42):6477–6482
12. Zhang R, Jiang B, Cao W (2001) Elastic, piezoelectric, and dielectric properties of multidomain 0.67Pb(Mg1/3Nb2/3)O3–0.33PbTiO3 single crystals. J Appl Phys 90:3471–3475
13. Zhang R, Jiang W, Jiang B, Cao W (2002) Elastic, dielectric and piezoelectric coefficients of domain engineered 0.70Pb(Mg1/3Nb2/3)O3–0.30PbTiO3 single crystal. In: Cohen RE (ed) Fundamental physics of ferroelectrics. American Institute of Physics, Melville, pp 188–197
14. Dantsiger AYa, Razumovskaya ON, Reznitchenko LA, Grineva LD, Devlikanova RU, Dudkina SI, Gavrilyatchenko SV, Dergunova NV, Klevtsov AN (1994) Highly effective piezoceramic materials (Handbook). Kniga, Rostov-on-Don (in Russian)
15. Gorish AV, Dudkevich VP, Kupriyanov MF, Panich AE, Turik AV (1999) Piezoelectric device-making. In: Physics of ferroelectric ceramics, vol 1. Radiotekhnika, Moscow (in Russian)
16. Kelly J, Leonard M, Tantigate C, Safari A (1997) Effect of composition on the electromechanical properties of $(1-x)Pb(Mg_{1/3}Nb_{2/3})O_3 - xPbTiO_3$ ceramics. J Am Ceram Soc 80:957–964

17. Gururaja TR, Safari A, Newnham RE, Cross LE (1988) Piezoelectric ceramic-polymer composites for transducer applications. In: Levinson LM (ed) Electronic ceramics: properties, devices, and applications. Marcel Dekker, Basel, pp 92–128
18. Topolov VYu, Bowen CR (2009) Electromechanical properties in composites based on ferroelectrics. Springer, London
19. Topolov VYu, Bowen CR (2015) High-performance 1–3-type lead-free piezo-composites with auxetic polyethylene matrices. Mater Lett 142:265–268
20. Zheng L, Huo X, Wang R, Wang J, Jiang W, Cao W (2013) Large size lead-free (Na, K)(Nb, Ta)O3 piezoelectric single crystal: growth and full tensor properties. CrystEngComm 15:7718–7722
21. Huo X, Zheng L, Zhang R, Wang R, Wang J, Sang S, Wang Y, Yang B, Cao W (2014) A high quality lead-free (Li, Ta) modified (K, Na)NbO3 single crystal and its complete set of elastic, dielectric and piezoelectric coefficients with macroscopic *4mm* symmetry. CrystEngComm 16:9828–9833
22. Evans KE, Alderson KL (1992) The static and dynamic moduli of auxetic microporous polyethylene. J Mater Sci Lett 11:1721–1724
23. Groznov IN (1983) Dielectric permittivity. In: Physics encyclopaedia. Sovetskaya Entsiklopediya, Moscow, pp 178–179 (in Russian)
24. Smotrakov VG, Eremkin VV, Alyoshin VA, Tsikhotsky ES (2000) Preparation and study of single-crystal–ceramic composite. Izv Akad Nauk, Ser Fiz 64:1220–1223 (in Russian)
25. Topolov VYu, Glushanin SV (2002) Effective electromechanical properties of ferroelectric piezoactive composites of the crystal–ceramic type based on $(Pb_{1-x}Ca_x)TiO_3$. Tech Phys Lett 28:279–282
26. Glushanin SV, Topolov VYu, Turik AV (2003) Prediction of piezoelectric properties of crystal-ceramic composites. Crystallogr Rep 48:491–499
27. Takahashi H, Tukamoto S, Qiu J, Tani J, Sukigara T (2003) Property of composite ceramics composed of single crystals and ceramic matrix using hybrid sintering. Jpn J Appl Phys 1 (42):6055–6058
28. Ikegami S, Ueda I, Nagata T (1971) Electromechanical properties of $PbTiO_3$ ceramics containing La and Mn. J Acoust Soc Am 50:1060–1066
29. Nagatsuma K, Ito Y, Jyomura S, Takeuchi H, Ashida S (1985) Elastic properties of modified $PbTiO_3$ ceramics with zero temperature coefficients. In: Taylor GW (ed) Ferroelectricity and related phenomena, vol 4., PiezoelectricityGordon and Breach Science Publishers, New York, pp 167–176
30. Levassort F, Thi MP, Hemery H, Marechal P, Tran-Huu-Hue L-P, Lethiecq M (2006) Piezoelectric textured ceramics: effective properties and application to ultrasonic transducers. Ultrasonics 44:e621–e626
31. Gibiansky LV, Torquato S (1997) On the use of homogenization theory to design optimal piezocomposites for hydrophone applications. J Mech Phys Solids 45:689–708
32. Sigmund O, Torquato S, Aksay IA (1998) On the design of 1–3 piezocomposites using topology optimization. J Mater Res 13:1038–1048
33. Topolov VYu, Filippov SE, Panich AE, Panich EA (2014) Highly anisotropic 1–3–0 composites based on ferroelectric ceramics: microgeometry—volume-fraction relations. Ferroelectrics 460:123–137
34. Asakawa K, Hiraoka T, Akasaka Y, Hotta Y (2003) Method for manufacturing porous structure and method for forming pattern. US Patent 6,565,763
35. Clark P, Moya W, (2003) Three dimensional patterned porous structures. US Patent 6,627,291
36. Ritter T, Geng X, Shung KK, Lopath PD, Park S-E, Shrout TR (2000) Single crystal PZN/PT —polymer composites for ultrasound transducer applications. IEEE Trans Ultrason Ferroelectr Freq Control 47:792–800
37. Cheng KC, Chan HLW, Choy CL, Yin Q, Luo H, Yin Z (2003) Single crystal PMN–0.33PT/epoxy 1–3 composites for ultrasonic transducer applications. IEEE Trans Ultrason Ferroelectr Freq Control 50:1177–1183

38. Ren K, Liu Y, Geng X, Hofmann HF, Zhang QM (2006) Single crystal PMN–PT/epoxy 1–3 composite for energy-harvesting application. IEEE Trans Ultrason Ferroelectr Freq Control 53:631–638
39. Wang F, He C, Tang Y, Zhao X, Luo H (2007) Single-crystal 0.7Pb(Mg1/3Nb2/3)O3–0.3PbTiO3/epoxy 1–3 piezoelectric composites prepared by the lamination technique. Mater Chem Phys 105:273–277
40. Cao H, Hugo Schmidt V, Zhang R, Cao W, Luo H (2003) Elastic, piezoelectric, and dielectric properties of $0.58Pb(Mg_{1/3}Nb_{2/3})O_3$–$0.42PbTiO_3$ single crystal. J Appl Phys 96:549–554
41. Topolov VYu, Krivoruchko AV, Bisegna P (2011) Electromechanical coupling and its anisotropy in a novel 1–3–0 composite based on single-domain $0.58Pb(Mg_{1/3}Nb_{2/3})O_3$–$0.42PbTiO_3$ crystal. Compos Sci Technol 71:1082–1088
42. Choy SH, Chan HLW, Ng MW, Liu PCK (2004) Study of 1–3 PZT fibre/epoxy composites with low volume fraction of ceramics. Integr Ferroelectr 63:109–115

Chapter 3
Figures of Merit of Modern Piezo-Active Ceramics and Composites

In the past decades, various figures of merit have been introduced to characterise the effectiveness of modern functional materials in the context of their piezoelectric and/or pyroelectric properties [1–5]. Knowledge of the figures of merit is important because of the links to various forms of energy and its transformation for exploitation of the functional materials as active elements in electronic circuits and devices. In the case of piezoelectric transducers, a number of figure of merits have been put forward to take into account the oscillation mode, the nature of the external loading, and the output conditions [see (1.30)–(1.32)]. For instance, piezoelectric transducers, transmit-receive systems and pulse-echo systems for broad-band airborne ultrasound applications can be described in terms of the figure of merit [7] that is regarded as a ratio of the electric power measured at the receiver output to the electric input power applied to the transmitter. Such a parameter is of value when dealing with a composite transducer that comprises of components with contrasting electromechanical properties. As follows from various literature data, various figures of merit suitable for piezoelectric materials are concerned with their piezoelectric coefficients, ECFs, specific acoustic impedance, quality factor, etc. In hydroacoustic applications, the relevant figure of merit strongly depends on the hydrostatic piezoelectric response of the active element [1, 7, 8]. As noted by Sessler and Hillenbrand [3], the choice of the figure of merit depends on the application for which the piezoelectric materials being used.

In this chapter we discuss examples of the figures of merit of various piezoelectric materials, that may be suitable for energy-harvesting applications, and links between the figures of merit and piezoelectric performance of these materials.

3.1 Poled Ferroelectric Ceramics

Values of squared figures of merit from (1.30)–(1.32) can vary in relatively wide ranges (Table 3.1). We arrange the perovskite-type poled FCs on $(Q_{33})^2$ that is concerned with the longitudinal piezoelectric effect. The largest $(Q_{33})^2$ values are

C.R. Bowen et al., *Modern Piezoelectric Energy-Harvesting Materials*,
Springer Series in Materials Science 238, DOI 10.1007/978-3-319-29143-7_3

achieved in the soft PCR-type FCs. The large $(Q_{33})^2$ value related to PZ 27 is accounted for by the relatively small dielectric permittivity $\varepsilon_{33}^{\sigma}$ in comparison to that of PCR-73 or PCR-7M. The relatively small $(Q_{33})^2$ value is achieved in the $PbTiO_3$-type FCs We remind the reader that the values of $(Q_{3j})^2$ in Table 3.1 are related to a low-frequency range [2–4] which is far from any resonance frequency of the piezoelectric element.

Data from Table 3.1 suggest that the inequality

$$(Q_{33})^2 \gg (Q_h)^2 \tag{3.1}$$

Table 3.1 Squared figures of merit $(Q_{3j})^2$ and $(Q_h)^2$ (in 10^{-12} Pa^{-1}) of perovskite-type FCs[a] at room temperature

FC	$(Q_{33})^2$	$(Q_{31})^2$	$(Q_h)^2$
PCR-73, hp[b]	13.9	2.72	0.188
PCR-7M, hp	13.1	2.77	0.0203
PCR-7, hp	12.0	2.53	0.0807
PZ 27	11.2	1.93	0.322
PZT-5	9.25	1.92	0.0724
PCR-1, hp	8.41	1.57	0.156
$Pb(Zr_{0.52}Ti_{0.48})O_3$	7.70	1.35	0.201
PZT-4	7.48	1.36	0.163
$(Pb_{0.94}Sr_{0.06})(Ti_{0.47}Zr_{0.53})O_3$	7.26	1.31	0.161
ZTS-19	7.10	1.20	0.228
PCR-8, ct	6.81	1.36	0.0782
PCR-8, hp	6.79	1.36	0.0726
PZT-7A	6.03	0.959	0.248
$Pb(Zr_{0.54}Ti_{0.46})O_3$	5.80	0.910	0.251
PCR-21, hp	5.23	0.994	0.0857
PZT-5H	4.50	0.743	0.158
$BaTiO_3$ (II)	2.40	0.404	0.0768
$BaTiO_3$ (I)	2.17	0.371	0.0648
PZ 34	2.01	0.00886	1.51
PCR-63, hp	1.89	0.348	0.0386
Modified $PbTiO_3$ (II)	1.87	0.0275	1.07
Modified $PbTiO_3$ (I)	1.66	0.0124	1.14
$(Pb_{0.855}Nd_{0.11})(Ti_{0.94}Mn_{0.02}In_{0.04})O_3$	1.53	0.0187	0.927
$(Pb_{0.9625}La_{0.025})(Ti_{0.99}Mn_{0.01})O_3$	1.50	0.0116	1.02
$(Pb_{0.85}Nd_{0.10})(Ti_{0.99}Mn_{0.01})O_3$	1.45	0.0132	0.947

[a]Full sets of electromechanical constants of the FCs are given in Table 1.2
[b]FC samples of the PCR type have been manufactured using either the conventional technology (ct) or hot pressing (hp). PCR is the abbreviation for the group "piezoelectric ceramics from Rostov-on-Don" (Russia) [7, 9, 10]

is not valid in the $PbTiO_3$-type FCs due to their large piezoelectric anisotropy The large piezoelectric anisotropy promotes a validity of the condition

$$(Q_{33})^2 \gg (Q_{31})^2 \tag{3.2}$$

irrespective of dielectric properties of the $PbTiO_3$-type FCs. It is noted that the poled FC is characterised by ∞mm symmetry [11, 12], and therefore, the equality $(Q_{31})^2 = (Q_{32})^2$ holds in the case of the poling axis OX_3.

Our numerous data show that the values of $(Q_{3j})^2$ and $(Q_h)^2$ from Table 3.1 remain relatively small in comparison to those achieved in various composites based on either FCs or relaxor-ferroelectric domain-engineered SCs [7, 10]. In Sects. 3.2–3.5 we discuss examples of large $(Q_{3j})^2$ and $(Q_h)^2$ for a number of connectivity patterns.

3.2 2–2 Composites

3.2.1 Ceramic/Polymer Composites

The 2–2 composite [7, 8, 10] is characterised by a continuous distribution of each component along two axes of a rectangular co-ordinate system $(X_1X_2X_3)$. Such a composite can be regarded as a laminar structure wherein the components (layers) are alternating along the third axis of $(X_1X_2X_3)$. The relative simplicity of the laminar structure and the possibility of varying the volume fractions of the components in a wide range allow the manufacture and design of 2–2 composites with tailored (e.g., extreme) values of their effective parameters. Methods to manufacture the 2–2 FC/polymer composites are discussed in review papers (see, for instance, [13–15]).

The 2–2 FC/polymer composites are usually manufactured in two modifications: with a series connection of adjacent layers (Fig. 3.1a) and with a parallel connection of adjacent layers (Fig. 3.1b). In the series-connected composite, the C1 and C2 layers are alternating in the OX_3 direction (Fig. 3.1a) and interfaces are perpendicular to OX_3, and this co-ordinate axis is the poling axis. In the parallel-connected composite, the poling axis OX_3 is parallel to the interfaces that separate the C1 and C2 layers (Fig. 3.1b). It is obvious that these two modifications are characterised by different electromechanical properties [7, 10] and anisotropy because of different orientations of the interfaces with respect to the poling axis OX_3. In the series-connected composite, the piezoelectric coefficients d_{3j}^* concerned with the poling axis obey the condition $d_{31}^* = d_{32}^* \neq d_{33}^*$. The piezoelectric coefficients d_{3j}^* of the parallel-connected composite satisfy the condition $d_{31}^* \neq d_{32}^* \neq d_{33}^*$ [7]. Based on these conditions, we state that the hydrostatic piezoelectric coefficient of the FC/polymer composite is represented in the general form as

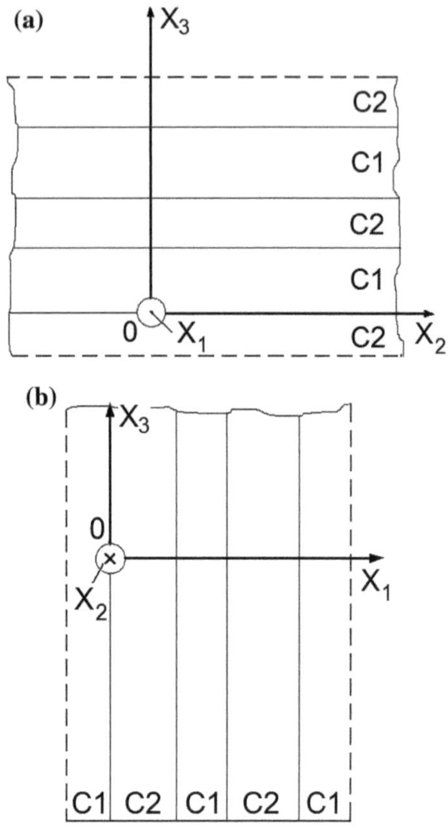

Fig. 3.1 Schematic of the 2–2 series-connected (**a**) and parallel-connected (**b**) composites with components denoted as C1 and C2. ($X_1X_2X_3$) is the rectangular co-ordinate system, and OX_3 is the poling direction (reprinted from monograph by Topolov et al. [10], with permission from Springer)

$$d_h^* = d_{31}^* + d_{32}^* + d_{33}^*. \tag{3.3}$$

Equation (3.3) is written on assumption that electrodes applied to the composite sample are perpendicular to the poling axis OX_3.

The effective electromechanical properties of the 2–2 composite are determined within the framework of the matrix approach [7, 10]. This approach is suitable for components with arbitrary symmetry of their properties at arbitrary volume fractions of the C1 and C2 components (Fig. 3.1). We assume that m is the volume fraction of the C1 component, and $1 - m$ is the volume fraction of the C2 component. The matrix of the effective electromechanical properties of the composite in the co-ordinate system ($X_1X_2X_3$) is written in the general form as

$$\|C^*\| = \begin{pmatrix} \|s^{*E}\| & \|d^*\|^t \\ \|d^*\| & \|\varepsilon^{*\sigma}\| \end{pmatrix}. \tag{3.4}$$

In (3.4),$\| s^{*E} \|$ is the matrix of elastic compliances at electric field E = const (6 × 6 matrix), $\| d^* \|$ is the matrix of piezoelectric coefficients (3 × 6 matrix), $\| \varepsilon^{*\sigma} \|$ is the matrix of dielectric permittivities at mechanical stress σ = const (3 × 3 matrix), and superscript 't' denotes the transposed matrix. The $\| C^* \|$ matrix from (3.4) is determined by averaging the electromechanical properties of the components on m and written as follows:

$$||C^*|| = [||C^{(1)}|| \cdot ||M||m + ||C^{(2)}||(1-m)] \cdot [||M||m + ||I||(1-m)]^{-1}. \quad (3.5)$$

In (3.5), $\| C^{(1)} \|$ and $\| C^{(2)} \|$ are matrices of the electromechanical properties of the FC (or C1) and polymer (or C2), respectively, $\| M \|$ is concerned with the electric and mechanical boundary conditions [7, 10] at interfaces (Fig. 3.1), and $\| I \|$ is the identity 9 × 9 matrix.

For example, in the case of the parallel-connected composite (Fig. 3.1b) with interfaces x_1 = const, the boundary conditions imply a continuity of components of mechanical stress $\sigma_{11} = \sigma_1$, $\sigma_{12} = \sigma_6$ and $\sigma_{13} = \sigma_5$, strain $\xi_{22} = \xi_2$, $\xi_{23} = \xi_4/2$ and $\xi_{33} = \xi_3$, electric displacement D_1, and electric field E_2 and E_3. The $\| M \|$ matrix from (3.5) is written for x_1 = const as $\| M \| = \| m_1 \|^{-1} \| m_2 \|$, where

$$||m_n|| = \begin{pmatrix} 1 & 0 & 0 & 0 & 0 & 0 & 0 & 0 & 0 \\ s_{12}^{(n),E} & s_{22}^{(n),E} & s_{23}^{(n),E} & s_{24}^{(n),E} & s_{25}^{(n),E} & s_{26}^{(n),E} & d_{12}^{(n)} & d_{22}^{(n)} & d_{32}^{(n)} \\ s_{13}^{(n),E} & s_{23}^{(n),E} & s_{33}^{(n),E} & s_{34}^{(n),E} & s_{35}^{(n),E} & s_{36}^{(n),E} & d_{13}^{(n)} & d_{23}^{(n)} & d_{33}^{(n)} \\ s_{14}^{(n),E} & s_{24}^{(n),E} & s_{34}^{(n),E} & s_{44}^{(n),E} & s_{45}^{(n),E} & s_{46}^{(n),E} & d_{14}^{(n)} & d_{24}^{(n)} & d_{34}^{(n)} \\ 0 & 0 & 0 & 0 & 1 & 0 & 0 & 0 & 0 \\ 0 & 0 & 0 & 0 & 0 & 1 & 0 & 0 & 0 \\ d_{11}^{(n)} & d_{12}^{(n)} & d_{13}^{(n)} & d_{14}^{(n)} & d_{15}^{(n)} & d_{16}^{(n)} & \varepsilon_{11}^{(n),\sigma} & \varepsilon_{12}^{(n),\sigma} & \varepsilon_{13}^{(n),\sigma} \\ 0 & 0 & 0 & 0 & 0 & 0 & 0 & 1 & 0 \\ 0 & 0 & 0 & 0 & 0 & 0 & 0 & 0 & 1 \end{pmatrix}$$

is represented using electromechanical constants of C1 (at n = 1) or C2 (at n = 2). We add that $\| m_n \|$ can be written taking into account the specific symmetry of each component. It should be noted that the effective electromechanical properties in (3.5) are determined in the longwave approximation [7], i.e., the wavelength of the external acoustic field is much greater than the width of each layer of the composite sample (Fig. 3.1).

Based on (3.4) and (3.5) and conventional formulae for the piezoelectric medium [11, 12], we obtain the effective parameters of the 2–2 composite at volume fractions of FC $0 < m < 1$. By analogy with (1.30)–(1.32), we represent squared figures of merit of the composite as

$$\left(Q_{3j}^*\right)^2 = d_{3j}^* g_{3j}^* \text{ and } \left(Q_h^*\right)^2 = d_h^* g_h^*, \quad (3.6)$$

where j = 1, 2 and 3, and

$$g_h^* = g_{31}^* + g_{32}^* + g_{33}^* \tag{3.7}$$

is the hydrostatic piezoelectric coefficient that characterises the sensitivity of the composite under hydrostatic loading.

Table 3.2 shows the volume-fraction dependence of the squared figures of merit $\left(Q_{3j}^*\right)^2$ of the 2–2 composite with a parallel connection of layers. This composite is based on the typical PZT-type FC. The parallel connection of the layers (Fig. 3.1b) that are continuous along OX_2 and OX_3 enables us to achieve large values of d_{33}^* and $\left|d_{32}^*\right|$ at relatively small volume fractions of FC m. A slow monotonic increase of $\varepsilon_{33}^{*\sigma}(m)$ and a rapid monotonic increase of $\left|d_{3j}^*(m)\right|$ at $m << 1$ lead to max$\left|g_{3j}^*(m)\right|$ and max$\left[Q_{3j}^*(m)\right]^2$ [7]. It should be added that $\left[Q_{33}^*(m)\right]^2$ undergoes greater changes at $0 < m < 1$ in comparison to $\left[Q_{31}^*(m)\right]^2$ and $\left[Q_{32}^*(m)\right]^2$ (see Table 3.2). This is due to the more pronounced piezoelectric activity of the composite along OX_3 where the considerable longitudinal piezoelectric effect of FC plays a dominating role.

Another example of the volume-fraction behaviour of $\left[Q_{3j}^*(m)\right]^2$ is given in Table 3.3. The performance of the 2–2 composite based on the highly anisotropic

Table 3.2 Volume-fraction dependences of the piezoelectric coefficients d_{3j}^* (in pC/N), g_{3j}^* (in mV m/N) and squared figures of merit $\left(Q_{3j}^*\right)^2$ (in 10^{-12} Pa^{-1}), which have been calculated for the 2–2 PZT-5 FC/araldite parallel-connected composite

m	d_{31}^*	d_{32}^*	d_{33}^*	g_{31}^*	g_{32}^*	g_{33}^*	$\left(Q_{31}^*\right)^2$	$\left(Q_{32}^*\right)^2$	$\left(Q_{33}^*\right)^2$
0	0	0	0	0	0	0	0	0	0
0.05	–43.7	–61.2	136	–79.4	–111	247	3.47	6.79	33.6
0.1	–67.0	–92.4	205	–55.7	–76.8	170	3.73	7.10	34.9
0.2	–93.1	–124	273	–35.2	–46.9	103	3.28	5.82	28.1
0.3	–109	–140	307	–26.3	–33.7	74.0	2.87	4.72	22.7
0.4	–121	–149	328	–21.3	–26.2	57.6	2.58	3.90	18.9
0.5	–131	–156	342	–18.1	–21.5	47.2	2.37	3.35	16.1
0.6	–140	–160	352	–15.9	–18.2	40.0	2.23	2.91	14.1
0.7	–148	–163	359	–14.3	–15.8	34.6	2.12	2.58	12.4
0.8	–155	–167	365	–13.1	–13.9	30.6	2.03	2.32	11.2
0.9	–163	–169	369	–12.1	–12.5	27.3	1.97	2.11	10.1
1	–170	–170	373	–11.3	–11.3	24.8	1.92	1.92	9.25

Note Extreme points of $g_{3j}^*(m)$ are min $g_{31}^* = g_{31}^*(0.016) = -102$ mV m/N, min $g_{32}^* = g_{32}^*(0.016) = -143$ mV m/N and max $g_{33}^* = g_{33}^*(0.015) = 320$ mV m/N

Table 3.3 Volume-fraction dependences of the piezoelectric coefficients d^*_{3j} (in pC/N), g^*_{3j} (in mV m/N) and squared figures of merit $\left(Q^*_{3j}\right)^2$ (in 10^{-12} Pa^{-1}), which have been calculated for the 2–2 modified $PbTiO_3$ (I) FC/araldite parallel-connected composite

m	d^*_{31}	d^*_{32}	d^*_{33}	g^*_{31}	g^*_{32}	g^*_{33}	$(Q^*_{31})^2$	$(Q^*_{32})^2$	$(Q^*_{33})^2$
0	0	0	0	0	0	0	0	0	0
0.05	−12.3	−5.22	28.8	−116	−49.1	270	1.43	0.256	7.78
0.1	−15.9	−5.39	37.2	−88.0	−29.8	205	1.40	0.161	7.63
0.2	−17.6	−5.11	43.8	−52.8	−15.3	131	0.929	0.0781	5.74
0.3	−17.1	−4.89	46.5	−35.1	−10.0	95.4	0.600	0.0489	4.44
0.4	−15.8	−4.75	48.0	−24.7	−7.40	74.9	0.390	0.0329	3.60
0.5	−14.2	−4.64	49.0	−17.8	−5.84	61.6	0.253	0.0271	3.02
0.6	−12.4	−4.57	49.6	−13.1	−4.81	52.3	0.159	0.0220	2.59
0.7	−10.5	−4.51	50.1	−9.50	−4.09	45.4	0.0998	0.0184	2.27
0.8	−8.50	−4.47	50.5	−6.76	−3.55	40.1	0.0575	0.0159	2.03
0.9	−6.47	−4.43	50.8	−4.58	−3.14	35.9	0.0296	0.0139	1.82
1	−4.40	−4.40	51.0	−2.81	−2.81	32.6	0.0124	0.0124	1.66

Note Extreme points of $d^*_{3j}(m)$ and $g^*_{3j}(m)$ are listed as follows: min $d^*_{31} = d^*_{31}(0.207) = -17.6$ pC/N, min $d^*_{32} = d^*_{32}(0.085) = -5.41$ pC/N, min $g^*_{31} = g^*_{31}(0.034) = -121$ mV m/N, min $g^*_{32} = g^*_{32}(0.021) = -62.8$ mV m/N, and max $g^*_{33} = g^*_{33}(0.031) = 285$ mV m/N

$PbTiO_3$-type FC suggests that max$\left[Q^*_{3j}(m)\right]^2$ are smaller than those in the 2–2 composite based on PZT-5 (see Table 3.2). The squared figure of merit $\left[Q^*_{33}(m)\right]^2$ of the 2–2 composite based on the $PbTiO_3$-type FC undergoes changes in a narrower range. This behaviour is a result of the lower piezoelectric activity of the $PbTiO_3$-type FC in comparison to the piezoelectric activity of the PZT-type FC. For instance, the piezoelectric coefficient $d^{(1)}_{33}$ of the modified $PbTiO_3$ (I) FC is approximately 7.3 times less than $d^{(1)}_{33}$ of the PZT-5 FC (see Table 3.2). The difference between values of $d^{(1)}_{31}$ of the aforementioned FCs is more pronounced. Thus, the 2–2 composite based on the PZT-type FC becomes preferable in the context of its squared figures of merit $\left[Q^*_{3j}(m)\right]^2 \gg \left(Q^{(1)}_{3j}\right)^2$.

We add that the series connection of the composite layers (Fig. 3.1a) lead to a low piezoelectric activity and small values of $\left[Q_{3j}^{*}(m)\right]^{2}$ due to interfaces that weaken the longitudinal piezoelectric effect despite the large $d_{33}^{(1)}$ value of the FC.

In general, the 2–2 FC/composite with a parallel connection of layers (Fig. 3.1b) is of value as a piezoelectric anisotropic material with squared figures of merit $\left[Q_{3j}^{*}(m)\right]^{2}$ that vary in wide ranges, and maxima of these parameters can be achieved at volume fractions of FC $m \approx 0.1$.

3.2.2 Single Crystal/Polymer Composites

Domain-engineered perovskite-type relaxor-ferroelectric PMN–*x*PT SCs with compositions near the morphotropic phase boundary [15–18] exhibit outstanding electromechanical properties that are of value for various piezotechnical applications. The high piezoelectric performance of the PMN–*x*PT SC/polymer composites [7, 10, 19–21] is important for such applications as active elements of energy harvesters, sensors, actuators, transducers, hydrophones, and acoustic antennae. A potential mechanism for improving the figures of merit is related to optimising the orientation of the main crystallographic axes [10] in the SC component.

The first example to be discussed is concerned with the parallel-connected 2–2 composite based on PMN–0.28PT SC. This composition is located close to the morphotropic phase boundary and can be regarded as a good candidate for the SC component in modern piezo-active composites. The PMN–0.28PT SC has been the first perovskite-type relaxor-ferroelectric SC, for which self-consistent and complete sets of electromechanical constants (Table 3.4) have been measured along three poling directions, namely, along [111], [001] and [011] in the perovskite unit cell [22]. As follows from experimental room-temperature data [22], the single-domain PMN–0.28PT SC is characterised by 3*m* symmetry with a spontaneous polarisation vector $\boldsymbol{P}_s^{(1)} \parallel$ [111]. The domain-engineered PMN–0.28PT SC poled along [001] exhibits 4*mm* symmetry and has an average spontaneous polarisation vector $\boldsymbol{P}_s^{(1)} \parallel$ [001]. The domain-engineered PMN–0.28PT SC poled along [011] leads to the domain-engineered state with *mm*2 symmetry and an average spontaneous polarisation vector $\boldsymbol{P}_s^{(1)} \parallel$ [011]. Liu et al. [22] underlined that the complete sets of electromechanical constants for the three crystallographic orientations (or poling directions) were determined from the same PMN–0.28PT crystal wafer, so that the composition and properties are uniform. It is seen from Table 3.4 that different poling directions lead to the different piezoelectric activity and anisotropy of the properties in PMN–0.28PT SCs. In this connection it is important to show advantages of the specific crystallographic orientation of the main crystallographic axes of the SC component at forming the piezoelectric response and figures of merit of the 2–2 SC/polymer composite.

Table 3.4 Elastic compliances $s_{ab}^{(n),E}$ (in 10^{-12} Pa^{-1}), piezoelectric coefficients $d_{ij}^{(n)}$ (in pC/N) and relative dielectric permittivity $\varepsilon_{pp}^{(n),\sigma}/\varepsilon 0$ of the poled PMN–0.28PT SC at room temperature [22]. Data on the [001]- and [011]-poled SCs are given with respect to the main crystallographic axes

Electromechanical constant	Single-domain [111]-poled SC, 3*m* symmetry	Polydomain [001]-poled SC, 4*mm* symmetry	Polydomain [011]-poled SC, *mm*2 symmetry
$s_{11}^{(n),E}$	8.78	44.57	13.40
$s_{12}^{(n),E}$	–4.90	–28.91	–21.18
$s_{13}^{(n),E}$	–0.93	–13.91	12.67
$s_{14}^{(n),E}$	16.87	0	0
$s_{22}^{(n),E}$	8.78	44.57	54.36
$s_{23}^{(n),E}$	–0.93	–13.91	–33.59
$s_{33}^{(n),E}$	6.32	34.38	28.02
$s_{44}^{(n),E}$	138.69	15.22	15.22
$s_{55}^{(n),E}$	138.69	15.22	147.06
$s_{66}^{(n),E}$	27.4	16.34	22.47
$d_{15}^{(n)}$	2382	122	2162
$d_{22}^{(n)}$	–312	0	0
$d_{24}^{(n)}$	2382	122	160
$d_{31}^{(n)}$	–43	–569	447
$d_{32}^{(n)}$	–43	–569	–1150
$d_{33}^{(n)}$	97	1182	860
$\varepsilon_{11}^{(n),\sigma}/\varepsilon_0$	4983	1672	4235
$\varepsilon_{22}^{(n),\sigma}/\varepsilon_0$	4983	1672	1081
$\varepsilon_{33}^{(n),\sigma}/\varepsilon_0$	593	5479	3873

Now we consider the 2–2 parallel-connected SC/polymer composite wherein a regular distribution of layers on the OX_1 direction is observed (Fig. 3.2). The SC and polymer layers are distributed continuously on the OX_2 and OX_3 directions. An orientation of the spontaneous polarisation vector $\boldsymbol{P}_s^{(1)}$ in each single-domain SC layer is characterised by the Euler angles φ, ψ and θ (Fig. 3.2, inset 1). The x, y and z axes are the main crystallographic axes in polydomain states (Fig. 3.2, insets 2 and 3). Interconnections between the spontaneous polarisation vectors $\boldsymbol{P}_{s,k}$ of several domain types and the spontaneous polarisation vector $\boldsymbol{P}_s^{(1)}$ of the polydomain SC layer are shown in insets 2 and 3 of Fig. 3.2. Hereby we note that $\boldsymbol{P}_s^{(1)} \parallel z$ in both cases. The polymer layers are ferroelectric with a remanent polarisation vector $\boldsymbol{P}_r^{(2)} \uparrow\downarrow OX_3$ (Fig. 3.2, inset 4). At this antiparallel orientation of $\boldsymbol{P}_r^{(2)}$ and OX_3, signs of the piezoelectric coefficients $d_{ij}^{(2)}$ of polyvinylidene fluoride (PVDF,

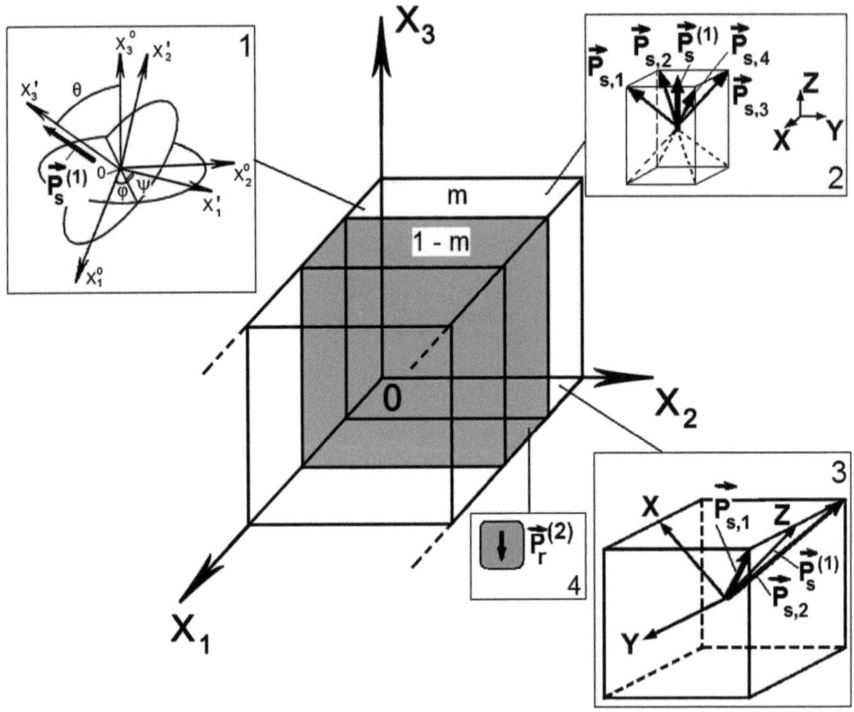

Fig. 3.2 Schematic of the 2–2 SC/polymer composite with parallel-connected layers. $(X_1X_2X_3)$ is a rectangular co-ordinate system, m and $1 - m$ are volume fractions of SC and polymer, respectively, $\boldsymbol{P}_s^{(1)}$ and $\boldsymbol{P}_r^{(2)}$ are the spontaneous polarisation vector of SC and the remanent polarisation vector of polymer, respectively. In *inset 1* φ, ψ and θ are Euler angles that characterise a rotation of the main crystallographic axes $\left(X_1^o X_2^o X_3^o\right) \rightarrow \left(X_1' X_2' X_3'\right)$ and the $\boldsymbol{P}_s^{(1)}$ vector in each single-domain [111]-poled SC layer. *Inset 2* comprises domain orientations in the [001]-poled SC with the effective spontaneous polarisation vector $\boldsymbol{P}_s^{(1)}$, *inset 3* comprises domain orientations in the [011]-poled SC with the effective spontaneous polarisation vector $\boldsymbol{P}_s^{(1)}$. $\boldsymbol{P}_{s,1}$, $\boldsymbol{P}_{s,2}$, $\boldsymbol{P}_{s,3}$, and $\boldsymbol{P}_{s,4}$ are spontaneous polarisation vectors of several domain types. x, y and z are main crystallographic axes of the polydomain SCs. The remanent polarisation vector of polymer $\boldsymbol{P}_r^{(2)} \uparrow\downarrow OX_3$ is shown in *inset 4* (reprinted from paper by Topolov et al. [27], with permission from Taylor and Francis)

Table 3.5) coincide [23] with sgnd_{ij} of conventional PZT-type FCs (see, for instance, Table 1.2). Manufacturing of the studied 2–2 composite with various orientations of the $\boldsymbol{P}_s^{(1)}$ and $\boldsymbol{P}_r^{(2)}$ vectors can be achieved by poling the ferroelectric components separately. As follows from literature data, the coercive fields $E_c^{(n)}$ of the single-domain PMN–xPT SCs [24, 25] and PVDF [26] obey the condition $E_c^{(1)} \ll E_c^{(2)}$ which favours the potential to individually pole the relaxor-ferroelectric SC and the ferroelectric polymer components.

Table 3.5 Elastic compliances $s_{ab}^{(n),E}$ (in 10^{-12} Pa^{-1}), piezoelectric coefficients $d_{ij}^{(n)}$ (in pC/N) and relative dielectric permittivity $\varepsilon_{pp}^{(n),\sigma}/\varepsilon_0$ of poled PVDF with $\boldsymbol{P}_r^{(2)} \uparrow\downarrow OX_3$ (∞mm symmetry) at room temperature [23]

Electromechanical constant	PVDF
$s_{11}^{(n),E}$	333
$s_{12}^{(n),E}$	–148
$s_{13}^{(n),E}$	–87.5
$s_{33}^{(n),E}$	299
$s_{44}^{(n),E}$	$1.90 \cdot 10^4$
$s_{66}^{(n),E}$	943
$d_{15}^{(n)}$	38
$d_{31}^{(n)}$	–10.42
$d_{33}^{(n)}$	33.64
$\varepsilon_{11}^{(n),\sigma}/\varepsilon_0$	7.513
$\varepsilon_{22}^{(n),\sigma}/\varepsilon_0$	7.513
$\varepsilon_{33}^{(n),\sigma}/\varepsilon_0$	8.431

Hereafter we distinguish the SC/polymer composites [27] with the following SC layers:

(i) Composite-1 with the single-domain layers (see Fig. 3.2, inset 1),
(ii) Composite-2 with the polydomain [001]-poled layers (see Fig. 3.2, inset 2) and
(iii) Composite-3 with the polydomain [011]-poled layers (see Fig. 3.2, inset 3).

The single-domain state in the SC layers of Composite-1 is stabilised by a bias field applied along the $\boldsymbol{P}_s^{(1)}$ vector. Rotations of the co-ordinate axes $(X_1^o X_2^o X_3^o) \rightarrow (X_1' X_2' X_3')$ (Fig. 3.2, inset 1) mean the rotations of the spontaneous polarisation vector $\boldsymbol{P}_s^{(1)}$ and the crystallographic axes of the single-domain SC layer. The co-ordinate axes OX_k^o and OX_k' (k = 1, 2 or 3) coincide at $\varphi = \psi = \theta = 0°$. In Composite-2 and Composite-3, where the SC layer comprises different domain types, we consider the rotation of the main crystallographic axes x, y and z of the SC layer around one of the co-ordinate axes, namely, $OX_1 \parallel$ x, $OX_2 \parallel$ y or $OX_3 \parallel$ z (Fig. 3.3). It is assumed that at these rotations, the spontaneous polarisation vectors of individual domains $\boldsymbol{P}_{s,1}$, $\boldsymbol{P}_{s,2}$ etc. in the SC layer are situated either over or in the $(X_1 O X_2)$ plane (see Fig. 3.2). In this case the rotation angles are varied in the following ranges:

(i) for Composite-1, $0° \le \varphi \le 120°$ (due to $3m$ symmetry of the single-domain SC, the upper limit is taken at 120° instead of 360°), $0° \le \psi \le 360°$ and $0° \le \theta \le 180°$ (Fig. 3.2, inset 1),
(ii) for Composite-2, $-45° \le \delta_x \le 45°$ and $-45° \le \delta_y \le 45°$ (Fig. 3.3a), and
(iii) for Composite-3, $-\arcsin(1/\sqrt{3}) \le \alpha \le \arcsin(1/\sqrt{3})$, $-45° \le \beta \le 45°$ and $0° \le \gamma \le 360°$ (see Fig. 3.3b and inset 3 in Fig. 3.2).

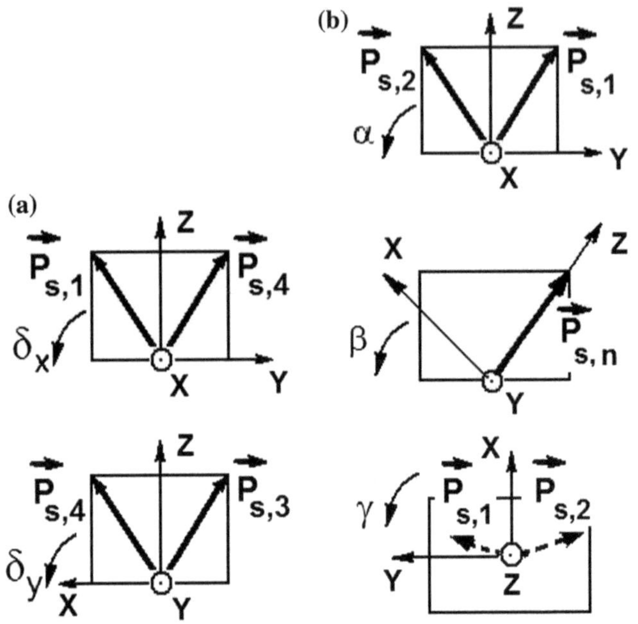

Fig. 3.3 Modes of rotation of the spontaneous polarisation vectors $P_{s,1}$, $P_{s,2}$, … in the [001]-poled (**a**) and [011]-poled (**b**) SCs. x, y and z are main crystallographic axes of the polydomain SCs, δ_x and δ_y are rotation angles for the [001]-poled SC, and α, β and γ are rotation angles for the [011]-poled SC (reprinted from paper by Topolov et al. [27], with permission from Taylor and Francis)

Because of the 4*mm* symmetry of the [001]-poled polydomain SC (see inset 2 in Fig. 1) and the equality of its piezoelectric coefficients $d_{31}^{(1)} = d_{32}^{(1)}$, we do not consider the rotation of the main crystallographic axes x, y and z around $OX_3 \parallel$ z in Composite-2.

The effective electromechanical properties and related parameters of the 2–2 SC/polymer composite are evaluated using (3.5) and complete sets of electromechanical constants of components (see Tables 3.4 and 3.5). Elements of the $\| C^* \|$ matrix from (3.5) are represented as functions of the volume fraction of SC m and rotation angles shown in Figs. 3.2 and 3.3.

By taking into account the symmetry features of the components (Tables 3.4 and 3.5) and the periodic laminar structure (Fig. 3.2), one can state the periodicity of the orientation dependence of the effective parameters of the 2–2 composite. For instance, its hydrostatic parameters obey equalities $\Pi_h^*(m, \varphi, \psi, \theta) = \Pi_h^*(m, 120° \pm \varphi, \psi, \theta)$ and $\Pi_h^*(m, \varphi, \psi, \theta) = \Pi_h^*(m, \varphi, 180° \pm \psi, \theta)$ for Composite-1, and $\Pi_h^*(m, \gamma) = \Pi_h^*(m, 180° \pm \gamma)$ for Composite-3 [27].

The squared hydrostatic figure of merit $(Q_h^*)^2$ from (3.6) strongly depends on the rotation mode of the main crystallographic axes (Table 3.6 and Fig. 3.3). Data from Table 3.6 enable us to conclude that Composite-3 is preferable due to the

Table 3.6 Absolute maxima of hydrostatic parameters Π_h* of the 2–2 PMN–0.28PT SC/PVDF composite at various rotation modes of the main crystallographic axes in the SC layers

Π_h*	Value of absolute maxΠ_h*	Volume fraction m and orientation angles which correspond to the value of maxΠ_h*
Composite-1		
d_h^*	133 pC/N	$m = 0.238$, $\varphi = 0°$, $\psi = 0°$, and $\theta = 74°$
g_h^*	242 mV m/N	$m = 0.007$, $\varphi = 60°$, $\psi = 90°$, and $\theta = 57°$
$(Q_h^*)^2$	$12.3 \cdot 10^{-12}$ Pa^{-1}	$m = 0.021$, $\varphi = 60°$, $\psi = 90°$, and $\theta = 63°$
Composite-2		
d_h^*	228 pC/N	$m = 0.253$ and $\delta_x = 0°$, or $m = 0.253$ and $\delta_y = 0°$
g_h^*	230 mV m/N	$m = 0.007$ and $\delta_x = 0°$, or $m = 0.007$ and $\delta_y = 0°$
$(Q_h^*)^2$	$15.3 \cdot 10^{-12}$ Pa^{-1}	$m = 0.038$ and $\delta_x = 0°$, or $m = 0.038$ and $\delta_y = 0°$
Composite-3		
d_h^*	372 pC/N	$m = 0.337$ and $\gamma = 90°$
g_h^*	284 mV m/N	$m = 0.012$ and $\gamma = 90°$
$(Q_h^*)^2$	$33.8 \cdot 10^{-12}$ Pa^{-1}	$m = 0.068$ and $\gamma = 90°$

largest maximum value of $(Q_h^*)^2$, the simplest orientation of the main crystallographic axes and the volume fraction of SC m that is more than 5% and can be achieved at manufacturing the composite sample. According to experimental data from Table 3.4, the squared hydrostatic figure of merit the [011]-poled PMN–0.28PT SC is $\left(Q_h^{(1)}\right)^2 = 0.719 \cdot 10^{-12}$ Pa^{-1}, i.e., by 47 times less than max$\left[(Q_h^*)^2\right]$ (Table 3.6) in Composite-3 based on this SC. Such a high hydrostatic parameter of Composite-3 is achieved due to the presence of the [011]-poled SC layers with an unusual combination of electromechanical constants that characterise the SC anisotropy. As seen from Table 3.4, the [011]-poled SC is characterised by elastic compliances that obey the untypical conditions $s_{11}^{(1),E} \approx s_{13}^{(1),E}$, $\left|s_{23}^{(1),E}\right| \approx 2.7\, s_{13}^{(1),E}$ and $s_{33}^{(1),E} \approx 2s_{13}^{(1),E}$, while the piezoelectric coefficients of this SC obey conditions $d_{33}^{(1)} \approx 1.9d_{31}^{(1)}$ and $\left|d_{32}^{(1)}\right| \approx 2.6d_{31}^{(1)}$. Variations of the rotation angles α and β (Fig. 3.3b) in the [011]-poled SC of Composite-3 would lead to values of maxΠ_h* which are lower than those listed in Table 3.6. This fact is also an argument for the important role of the anisotropy of the electromechanical properties of the SC component in forming the hydrostatic piezoelectric response of the 2–2 composite.

It should be added for comparison that values of max d_h^*, max g_h^* and max $\left[(Q_h^*)^2\right]$ of the studied composites (Table 3.6) highlight their advantages over the conventional 2–2 PZT-type FC/polymer composites, for which typical values are max $d_h^* \approx$ (50–80) pC/N and max $g_h^* \approx$ (100–300) mV m/N [28]. It is also known that experimental values of max $\left[(Q_h^*)^2\right] \approx 5 \cdot 10^{-11}$ Pa^{-1} are related to a 2–2 PZT-type FC/polymer composite [15] with a specific orientation of the layers with respect to the poling direction. Thus, the orientation effect in the piezo-active 2–2 composites plays the important role in achieving the large values of $(Q_h^*)^2$ that is important for hydroacoustic applications.

The second example to be discussed is concerned with a 2–2 SC/auxetic polymer composite. The auxetic polymer component is characterised by a negative Poisson's ratio, i.e., elastic compliances of the isotropic material are $s_{11}^{(n)} > 0$ and $s_{12}^{(n)} > 0$ (see Table 2.7)[1]. This elastic feature can influence the electromechanical properties and related parameters of the composite [29] in a wide volume-fraction range. Among the SC components, of interest are perovskite-type relaxor-ferroelectric xPb($In_{0.5}Nb_{0.5}$)·$O_3 - y$Pb($Mg_{1/3}Nb_{2/3}$)$O_3 - (1 - x - y)PbTiO_3$ (PIN-x-y) SCs in the single-domain state [30] at room temperature (see, for instance, data in Table 3.7). The full set of

Table 3.7 Room-temperature elastic compliances s_{ab}^E (in 10^{-12} Pa^{-1}), piezoelectric coefficients d_{ij} (in pC/N) and relative dielectric permittivity $\varepsilon_{pp}^{\sigma}/\varepsilon_0$ of the single-domain PIN-0.24-0.49 SC (3m symmetry) [30]

Electromechanical constant	Single-domain [111]-poled PIN-0.24-0.49 SC
$s_{11}^{(n),E}$	11.62
$s_{12}^{(n),E}$	−7.81
$s_{13}^{(n),E}$	−1.10
$s_{14}^{(n),E}$	22.17
$s_{22}^{(n),E}$	11.62
$s_{23}^{(n),E}$	−1.10
$s_{33}^{(n),E}$	6.05
$s_{44}^{(n),E}$	93.33
$s_{55}^{(n),E}$	93.33
$s_{66}^{(n),E}$	38.85
$d_{15}^{(n)}$	2015
$d_{22}^{(n)}$	−490
$d_{24}^{(n)}$	2015
$d_{31}^{(n)}$	−21
$d_{32}^{(n)}$	−21
$d_{33}^{(n)}$	75
$\varepsilon_{11}^{(n),\sigma}/\varepsilon_0$	5800
$\varepsilon_{22}^{(n),\sigma}/\varepsilon_0$	5800
$\varepsilon_{33}^{(n),\sigma}/\varepsilon_0$	650

[1]For auxetic polyethylene it is assumed that its dielectric permittivity equals 2.3 as is known for monolithic polyethylene samples [31].

electromechanical constants of the single-domain PIN-0.24-0.49 SC testifies to the obvious anisotropy of its properties.

It is assumed that the 2–2 single-domain PIN-0.24-0.49 SC/auxetic polyethylene is characterised by the regular arrangement of layers as shown in Fig. 3.2. We use the Euler angles φ, ψ and θ to describe the rotation of the main crystallographic axes $\left(X_1^o X_2^o X_3^o\right) \rightarrow \left(X_1' X_2' X_3'\right)$ and the orientation of the spontaneous polarisation vector $\boldsymbol{P}_s^{(1)}$ in each single-domain SC layer (see inset 1 in Fig. 3.2). In the studied 2–2 composite based on the single-domain PIN-0.24-0.49 SC, large values of absolute maxima of g_h^* and $\left(Q_h^*\right)^2$ are achieved at very low volume fractions of SC: absolute max $g_h^* = 3860$ mV m/N at $m = 0.0008$ and absolute max $\left[\left(Q_h^*\right)^2\right] = 1.71 \cdot 10^{-9}$ Pa^{-1} at $m = 0.0014$ [29]. This performance is accounted for by the large elastic compliances of the auxetic component (see Table 3.7) in comparison to the elastic compliances of the SC component and by the effect of the dielectric properties of the auxetic component on the non-monotonic behaviour of g_h^* and $\left(Q_h^*\right)^2$ at $m << 1$. It is obvious that the aforementioned values of $m < 0.01$ do not enable one to create a composite sample with the extremely large value of $\left(Q_h^*\right)^2$. To avoid this limitation, we should consider the performance of this composite at larger volume fractions of SC.

Examples of the behaviour of $\left(Q_h^*\right)^2$ in vicinities of local maxima achieved at variations of the Euler angles are shown in Fig. 3.4. The considerable decrease of $\left(Q_h^*\right)^2$ in comparison to the values of absolute max $\left[\left(Q_h^*\right)^2\right] = 1.71 \cdot 10^{-9}$ Pa^{-1} [29] is concerned with the increase of dielectric permittivity $\varepsilon_{33}^{*\sigma}$ that strongly influences

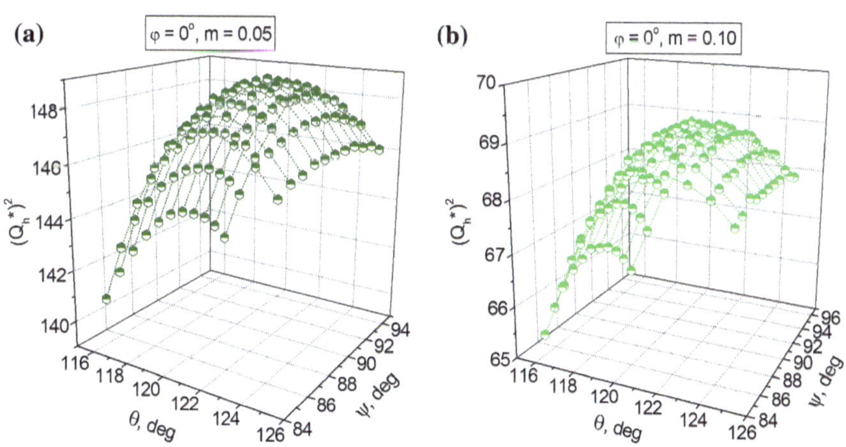

Fig. 3.4 Hydrostatic squared figure of merit $\left(Q_h^*\right)^2$ (in 10^{-12} Pa^{-1}) of the 2–2-type PIN-0.24-0.49/auxetic polyethylene composite in the vicinity of local max $\left[\left(Q_h^*\right)^2\right]$ at $m = 0.05$ (**a**) and $m = 0.10$ (**b**) (reprinted from paper by Topolov et al. [29], with permission from Wiley-VCH)

[7, 10] the piezoelectric coefficients g_{33}^* and g_h^* and, therefore, $(Q_h^*)^2$ in accordance with (3.6). At volume fractions of SC $0.05 \le m \le 0.10$, irrespective of the orientation of the main crystallographic axes in the SC layer, the inequality $\varepsilon_{33}^{*\sigma} \ll \varepsilon_{33}^{(1),\sigma}$ holds, and this feature of the composite favours relatively large values of $g_h^* \sim 10^2$ mV m/N and $(Q_h^*)^2 \sim 10^{-11}$–10^{-10} Pa^{-1} [29] which are of interest for hydroacoustic applications.

The third example of the performance is concerned with local maxima of squared figures of merit $(Q_{33}^*)^2$ and $(Q_h^*)^2$. These parameters were analysed for a 2–2-type polydomain PMN–*x*PT SC/auxetic polyethylene composite [32] with x = 0.28–0.33. As a result of the complicated microgeometry of the auxetic polymer component, we use the notation '2–2-type' instead of '2–2'. As follows from Fig. 3.5, the squared figures of merit $(Q_{33}^*)^2$ and $(Q_h^*)^2$ decrease with an increase in the volume fraction of SC m. However values of local maxima of both $(Q_{33}^*)^2$ and $(Q_h^*)^2$ at $0.05 < m < 0.2$ are still large in comparison to the similar values related to the conventional 2–2 FC/polymer composites [7, 14, 15]. The considerable anisotropy of the electromechanical properties of the SC component and the unusual elastic properties of the polymer component promote the anisotropy of figures of merit which correspond to the longitudinal and transverse piezoelectric responses of the composite. The inequality

$$(Q_{33}^*)^2/\left(Q_{3j}^*\right)^2 \ge 10\,(j = 1 \text{ and } 2) \tag{3.8}$$

holds [32] within wide ranges of m and rotation angles (Fig. 3.6) and at different rotation modes. The considerable part of the hatched area in Fig. 3.6a is related to the volume fraction m and orientation angle α that promote large values of $(Q_h^*)^2$.

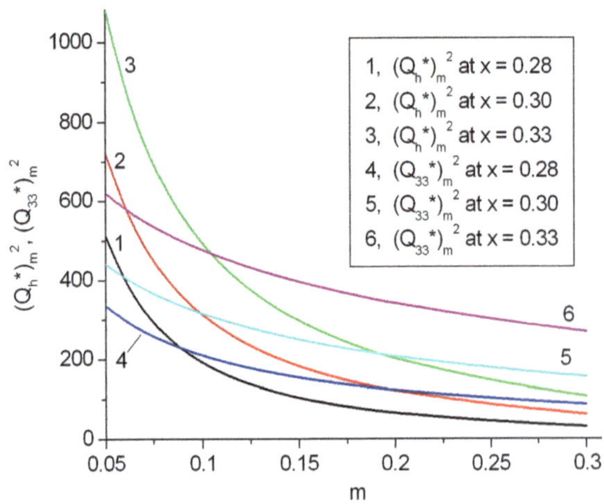

Fig. 3.5 Local maxima of squared figures of merit $(Q_h^*)m^2$ and $(Q_{33}^*)m^2$ (in 10^{-12} Pa^{-1}) of the 2–2-type PMN–*x*PT SC/auxetic polyethylene composite with SC layers poled along [001] (reprinted from paper by Bowen and Topolov [32], with permission from Taylor and Francis)

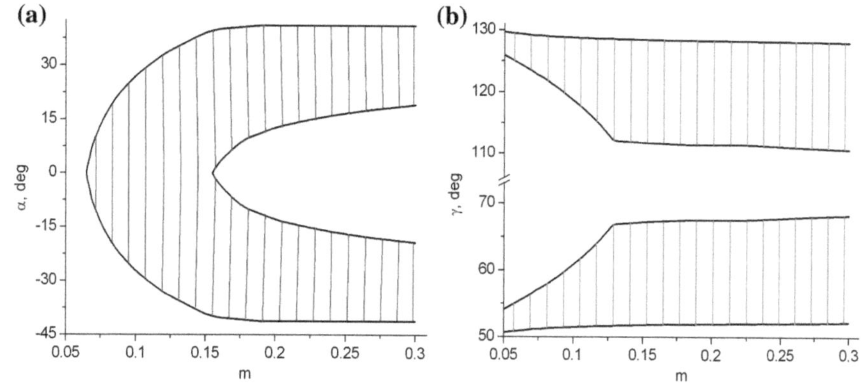

Fig. 3.6 Examples of validity of condition (3.8) (hatched areas in graphs) at different rotation modes in the 2–2-type PMN–xPT SC/auxetic PE composite: a, $x = 0.33$ and SC layers poled along [001], and b, $x = 0.29$ and SC layers poled along [011] (reprinted from paper by Bowen and Topolov [32], with permission from Taylor and Francis)

The rotation mode concerned with α is shown in Fig. 3.3b. We note that the validity of the condition (3.8) is linked with the appreciable orientation effect and elastic anisotropy of the studied composite and the relatively high piezoelectric activity of its SC component irrespective of its poling direction. It is clear that condition (3.8) is similar to (3.2) related to the $PbTiO_3$-type FCs with a large piezoelectric anisotropy. However the condition (3.8) was formulated [32] for 2–2 composites based on PMN–xPT SCs with large piezoelectric coefficients $d_{3j}^{(1)}$.

Thus, the results shown in Sect. 3.2 enable us to characterise the piezo-active 2–2 composites as materials with figures of merit that can vary in wide ranges. Of particular interest are the 2–2 composites based on relaxor-ferroelectric SCs that are poled along specific directions and promote the high piezoelectric activity and sensitivity of the composites.

3.3 1–3 Composites

In the 1–3 composite, the main piezo-active component is self-connected in one dimension (often along the poling axis) and the piezo-passive component (or the component with a low piezoelectric activity) is self-connected in three dimensions. In fact, it is a system of long aligned poled rods embedded in a large matrix. The 1–3 FC/polymer and relaxor-ferroelectric SC/polymer composites (Figs. 2.2 and 2.3) are widespread [7, 8, 10, 14, 15, 19, 20] due to the ease of poling and a variety of advantages over the poled monolithic piezo-active components. Often the FC component can be represented by a continuous row of particles, relatively long rods or fibres, or a series of discs. The cross-section of the FC rod can be in the form of a square, circle, ellipse, triangle, etc. The combination of the relatively low dielectric permittivity and high piezoelectric activity at small volume fractions of the FC or SC component results in a significant piezoelectric sensitivity of the 1–3 composite [7, 19, 20]. The 1–3 relaxor-ferroelectric SC/polymer composite is of interest due to the combination of the high piezoelectric activity and sensitivity, considerable electromechanical coupling, hydrostatic piezoelectric response, etc. Orientation effects studied in such a composite [10] represent an independent interest at improving the performance in comparison to the traditional FC/polymer composites. Undoubtedly, advantages of the 1–3 composites over individual FC or SC components stimulate intensive research into the development and manufacture of 1–3 and related piezo-active composites in the last decades. An additional stimulus is achieved by modification of the polymer matrix; see, for instance, [10, 33–35].

In Sect. 3.3 we analyse examples of figures of merit and ways to increase these parameters in the 1–3 and related composites based on either FCs or domain-engineered relaxor-ferroelectric SCs.

3.3.1 Ceramic/Polymer Composites

We assume that the 1–3 composite consists of long FC rods (component 1) that are regularly distributed in a continuous polymer matrix (component 2). The rods are considered to be in the form of circular cylinders aligned parallel to the poling OX_3 axis and continuous in the same direction (Fig. 3.7). In the (X_1OX_2) plane, a square arrangement of the cylinders is formed.

The effective electromechanical properties of the 1–3 composite are represented in the general form [7, 10] as

$$||C^*|| = ||C^{(2)}|| + m\left(||C^{(1)}|| - ||C^{(2)}||\right)||A||. \tag{3.9}$$

In (3.9), $||C^{(n)}|| = \begin{pmatrix} ||c^{(n),E}|| & ||e^{(n)}||^t \\ ||e^{(n)}|| & -||\varepsilon^{(n),\xi}|| \end{pmatrix}$ is the 9 × 9 matrix that characterises the electromechanical properties of the FC rod (at $n = 1$) and the surrounding polymer

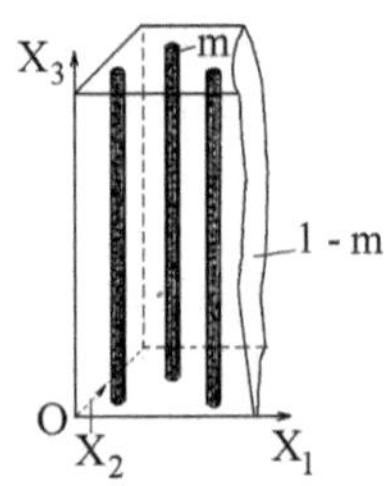

Fig. 3.7 Schematic of the 1–3 composite with cylindrical rods. m and $1 - m$ are volume fractions of the FC rods and the polymer matrix, respectively. $(X_1X_2X_3)$ is the rectangular co-ordinate system (reprinted from monograph by Topolov and Bowen [7], with permission from Springer)

matrix (at $n = 2$), is the mechanical strain—electric field concentration matrix (9 × 9 matrix), $\| c^{(n),E} \|$ is the 6 × 6 matrix of elastic moduli measured at constant electric field, $\| e^{(n)} \|$ is the 6 × 3 matrix of piezoelectric coefficients, $\| \varepsilon^{(n),\xi} \|$ is the 3 × 3 matrix of dielectric permittivities measured at mechanical strain ξ = const, and superscript 't' denotes the transposition. The $\| A \|$ matrix from (3.9) is related to the boundary conditions in the 'rod—surrounding medium' region. Taking into account an electromechanical interaction between the piezo-active rods in the composite and following the effective field method [7, 10, 36, 37], we write $\| A \|$ as

$$||A|| = \left[||I|| + (1-m)||S||\,||C^{(2)}||^{-1}\left(||C^{(1)}|| - ||C^{(2)}||\right)\right]^{-1}. \quad (3.10)$$

In (3.10), $\| I \|$ is the 9 × 9 identity matrix, and $\| S \|$ is the 9 × 9 matrix containing components of the Eshelby electroelastic tensor [37–39]. Elements of $\| C^* \|$ from (3.9) are functions of the volume fraction m. Taking into account the square arrangement of the rods the (X_1OX_2) plane (Fig. 3.7), we note that the upper limit for their volume fraction would be π/4. A transition from the full set of electromechanical constants, that constitute $\| C^* \|$, to the piezoelectric coefficients d_{ij}^*, g_{ij}^*, squared figures of merit from (3.6), etc. is carried out using conventional formulae [11, 12] for the piezoelectric medium.

As is known [7, 35], values of max $\left[Q_{33}^*(m)\right]^2$ and max $\left[Q_h^*(m)\right]^2$ in the 1–3 composite based on the PZT-type FC, are several tens of times larger than the $\left[Q_{33}^*(1)\right]^2$ and $\left[Q_h^*(1)\right]^2$ values related to the monolithic FC component. As a rule, max $\left[Q_{33}^*(m)\right]^2$ and max $\left[Q_h^*(m)\right]^2$ are attained at relatively small volume fractions of the PZT-type FC, i.e., at $0 < m < 0.1$ [7, 35]. In the same m range, the 1–3 composite is characterised by a large anisotropy of the piezoelectric coefficients e_{3j}^*, i.e., $e_{33}^*/|e_{31}^*| \gg 1$. Taking into account these results, we represent (3.6) in the form [35]

$$(Q_h^*)^2 \approx \eta_{elas,h}^*\left(e_{33}^*/c_{33}^{*E}\right)^2/\varepsilon_{33}^{*\sigma} \approx \eta_{elas,h}^*\left[m\,e_{33}^{(1)}/\left(mc_{33}^{(1),E} + c_{33}^{(2),E}\right)\right]^2/\left(m\,\varepsilon_{33}^{(1),\sigma}\right)$$
$$\text{and } \left(Q_{33}^*\right)^2 \approx \left(\eta_{elas,33}^*/\eta_{elas,h}^*\right)\left(Q_h^*\right)^2. \quad (3.11)$$

In (3.11), $\eta_{elas,h}^* = \left\{\left[1 + (\beta_{12}^*)^{-1} - 2(\beta_{13}^*)^{-1}\right]/\left[1 + (\beta_{12}^*)^{-1} - 2(\beta_{13}^*)^{-1}(\beta_{33}^*/\beta_{13}^*)\right]\right\}^2$, $\eta_{elas,33}^* = \left\{\left[1 + (\beta_{12}^*)^{-1}\right]/\left[1 + (\beta_{12}^*)^{-1} - 2(\beta_{13}^*)^{-1}(\beta_{33}^*/\beta_{13}^*)\right]\right\}^2$ and $\beta_{ab}^* = c_{11}^{*E}/c_{ab}^{*E}$ depend on the elastic properties of the composite only, and an inequality $e_{33}^{(1)} \gg e_{33}^{(2)}$ holds for its components. $(Q_h^*)^2$ and $(Q_{33}^*)^2$ from (3.11) depend on

$\gamma = me_{33}^{(1)} / \left(mc_{33}^{(1),E} + c_{33}^{(2),E} \right)$, so that a higher $e_{33}^{(1)} / c_{33}^{(1),E}$ ratio leads to larger $(Q_h^*)^2$ and $(Q_{33}^*)^2$ at $0 < m < 0.1$.

We note that among the perovskite-type FCs listed in Table 1.2 and C.1, PCR-7M is characterised by the highest $e_{33}^{(1)} / c_{33}^{(1),E}$ ratio at room temperature. For example, max $\left[Q_h^*(m)\right]^2 = 6.24 \cdot 10^{-12}$ Pa^{-1} and max $\left[Q_{33}^*(m)\right]^2 = 6.24 \cdot 10^{-10}$ Pa^{-1} are achieved in the 1–3 PCR-7M/elastomer composite [35]. The max $\left[Q_h^*(m)\right]^2$ value is approximately three times larger than that of typical 1–3 composites based on the PZT-type FCs [7, 14].

Data on the piezoelectric properties and squared figures of merit of the studied composites based on FCs are shown in Table 3.8. It is seen that values of $(Q_{3j}^*)^2$ of the PZT-5-based composite are comparable to $(Q_{3j}^*)^2$ of the 2–2 composite with the same components (see Table 3.8). The $(Pb_{0.75}Ca_{0.25})TiO_3$ FC belongs to the highly anisotropic $PbTiO_3$-type FCs [40] and does not give rise to large values of $(Q_{3j}^*)^2$ in the related 1–3 composite. The combination of the PCR-7M FC (i.e., soft ferroelectric material with large values of $d_{33}^{(1)}$ and $\left|d_{31}^{(1)}\right|$ [9]) and auxetic polyethylene leads to very large values of $(Q_{33}^*)^2$ at non-monotonic changes in $(Q_{31}^*)^2$. These changes are caused by decreasing d_{31}^* in the wide m range and by changing sgn d_{31}^* (see Table 3.8). The 1–3-type PCR-7M FC/auxetic polyethylene composite is also of interest due to the relatively large volume-fraction range wherein condition (3.8) for the large anisotropy of $(Q_{3j}^*)^2$ is valid. This is a result of the active role of the auxetic,namely, its unusual elastic properties, in forming the piezoelectric properties and their considerable anisotropy [33] in the studied composite.

3.3.2 *Single Crystal/Polymer Composites*

The 1–3 relaxor-ferroelectric SC/polymer composite is of interest for various piezotechnical applications, including piezoelectric energy harvesting [19, 20]. The SC component with large piezoelectric coefficients $d_{3j} \sim 10^3$ pC/N (see Table 1.1) favours the strong electromechanical coupling (see Sect. 2.5.2), high piezoelectric sensitivity and activity, etc. [7, 42]. Taking the domain-engineered SC component, one can form the structure of the SC rods that are regularly arranged in the continuous polymer matrix (Fig. 2.4) and strongly influence the piezoelectric properties in the composite system. By varying the volume fraction of the SC rods m, one can tailor a composite sample with effective parameters near their extreme points that are predicted by various methods [7]. Taking into account a rotation of the main crystallographic axes in each SC rod and the related orientation effect [10, 43], one can improve the piezoelectric response and related parameters of the 1–3 composite. A modification of the polymer matrix, for instance by forming a porous

Table 3.8 Volume-fraction dependences of the piezoelectric coefficients d^*_{3j} (in pC/N), g^*_{3j} (in mV m/N) and squared figures of merit $\left(Q^*_{3j}\right)^2$ (in 10^{-12} Pa^{-1}), which have been calculated for the 1–3-type composites with cylindrical rods

m	d^*_{31}	d^*_{33}	g^*_{31}	g^*_{33}	$(Q^*_{31})^2$	$(Q^*_{33})^2$
1–3 PZT-5 FC/araldite composite						
0	0	0	0	0	0	0
0.05	–52.8	141	–56.1	150	2.96	21.2
0.1	–79.6	210	–40.1	106	3.19	22.3
0.2	–108	277	–25.7	65.9	2.78	18.3
0.3	–124	310	–19.1	47.8	2.37	14.8
0.4	–135	330	–15.4	37.6	2.08	12.4
0.5	–143	343	–12.9	30.9	1.84	10.6
0.6	–150	353	–11.2	26.4	1.68	9.32
1–3 $(Pb_{0.75}Ca_{0.25})TiO_3$ FC/araldite composite [40]						
0	0	0	0	0	0	0
0.05	–7.28	19.9	–69.1	189	0.503	3.76
0.1	–6.73	22.0	–38.4	126	0.258	2.77
0.2	–6.65	25.0	–21.1	79.3	0.140	1.98
0.3	–5.99	26.1	–13.1	57.3	0.0785	1.50
0.4	–5.19	26.8	–8.69	44.9	0,0451	1.19
0.5	–4.35	27.2	–5.89	36.9	0.0256	0.893
0.6	–3.51	27.4	–3.99	31.2	0.0140	0.855
1–3-type PCR-7M FC/auxetic polyethylene composite [33]						
0	0	0	0	0	0	0
0.05	415	669	193	310	80.1	207
0.1	318	698	72.4	158	23.0	110
0.2	153	718	17.1	80.3	2.62	57.7
0.3	30.1	729	2.23	53.9	0.0671	39.3
0.4	–64.7	736	–3.57	40.5	0.231	29.8

(continued)

Table 3.8 (continued)

m	d_{31}^*	d_{33}^*	g_{31}^*	g_{33}^*	$(Q_{31}^*)^2$	$(Q_{33}^*)^2$
0.5	–141	741	–6.20	32.5	0.874	24.1
0.6	–206	745	–7.53	27.2	1.55	20.3

Notes
1. The full set of room-temperature electromechanical constants of the $(Pb_{0.75}Ca_{0.25})TiO_3$ FC was taken from work [41]. Elastic and dielectric constants of auxetic polyethylene were taken from Table 2.7
2. The effective field method was used to calculate the effective electromechanical properties of the 1–3 PZT-5 FC/araldite and $(Pb_{0.75}Ca_{0.25})TiO_3$ FC/araldite composites
3. The finite element method was used to calculate the effective electromechanical properties of the 1–3-type PCR-7M FC/auxetic polyethylene composite
4. For the $(Pb_{0.75}Ca_{0.25})TiO_3$ FC/araldite composite, differences between the effective parameters calculated using the effective field method and those calculated by means of the finite element method are less than 3 % [40]
5. We do not consider specifics of connectivity of the PCR-7M FC/auxetic polyethylene composite because of complicated microgeometry of the auxetic polymer matrix, and hereafter the notation '1–3-type' is used

structure therein [42] or incorporating FC inclusions therein [44–46], provides additional opportunities to improve composite performance.

Squared figures of merit $(Q_{33}^*)^2$ and $(Q_h^*)^2$ of the 1–3 composite based on the relaxor-ferroelectric domain-engineered [001]-poled SC vary in wide ranges (curve 1 in Fig. 3.8), however their maxima are observed at relatively small volume fractions on SC, $m \approx 0.1$–0.2. Values of $(Q_{33}^*)^2$ near its maximum (see curve 1 in Fig. 3.8a) are about 7–10 times more than $(Q_{33}^*)^2$ of the 1–3 PZT-5 FC/araldite composite and comparable to $(Q_{33}^*)^2$ of the 1–3-type PCR-7M FC/auxetic polyethylene composite (see Table 3.8). By replacing the monolithic polymer matrix with a porous polymer matrix, wherein the spherical air pores are distributed regularly, we achieve larger values of $(Q_{33}^*)^2$ and $(Q_h^*)^2$ without considerable shifts of their maximum points (curve 2 in Fig. 3.8). This is due to the active role of the elastic properties of the matrix at its dielectric permittivity $\varepsilon_{33}^{(2)} \ll \varepsilon_{33}^{*\sigma}$. Values of $(Q_h^*)^2 \approx 6 \cdot 10^{-12}$ Pa^{-1} (curve 1 in Fig. 3.8b) are more than $(Q_h^*)^2$ of a 1–3–0 PZT rod composite [14] and close to max $(Q_h^*)^2$ of the 1–3 PCR-7M/elastomer composite [35]. Values of $(Q_h^*)^2 \approx 25 \cdot 10^{-12}$ Pa^{-1} (curve 2 in Fig. 3.8b) exceed $(Q_h^*)^2$ of such FC-based materials as 1–3–0 PZT rods—foamed polymer and 3–2 perforated composites [14].

Large values of both $(Q_{33}^*)^2$ and $(Q_h^*)^2$ are achieved in the novel 1–3-type lead-free composites at relatively small volume fractions of SC m (Fig. 3.9). Such a

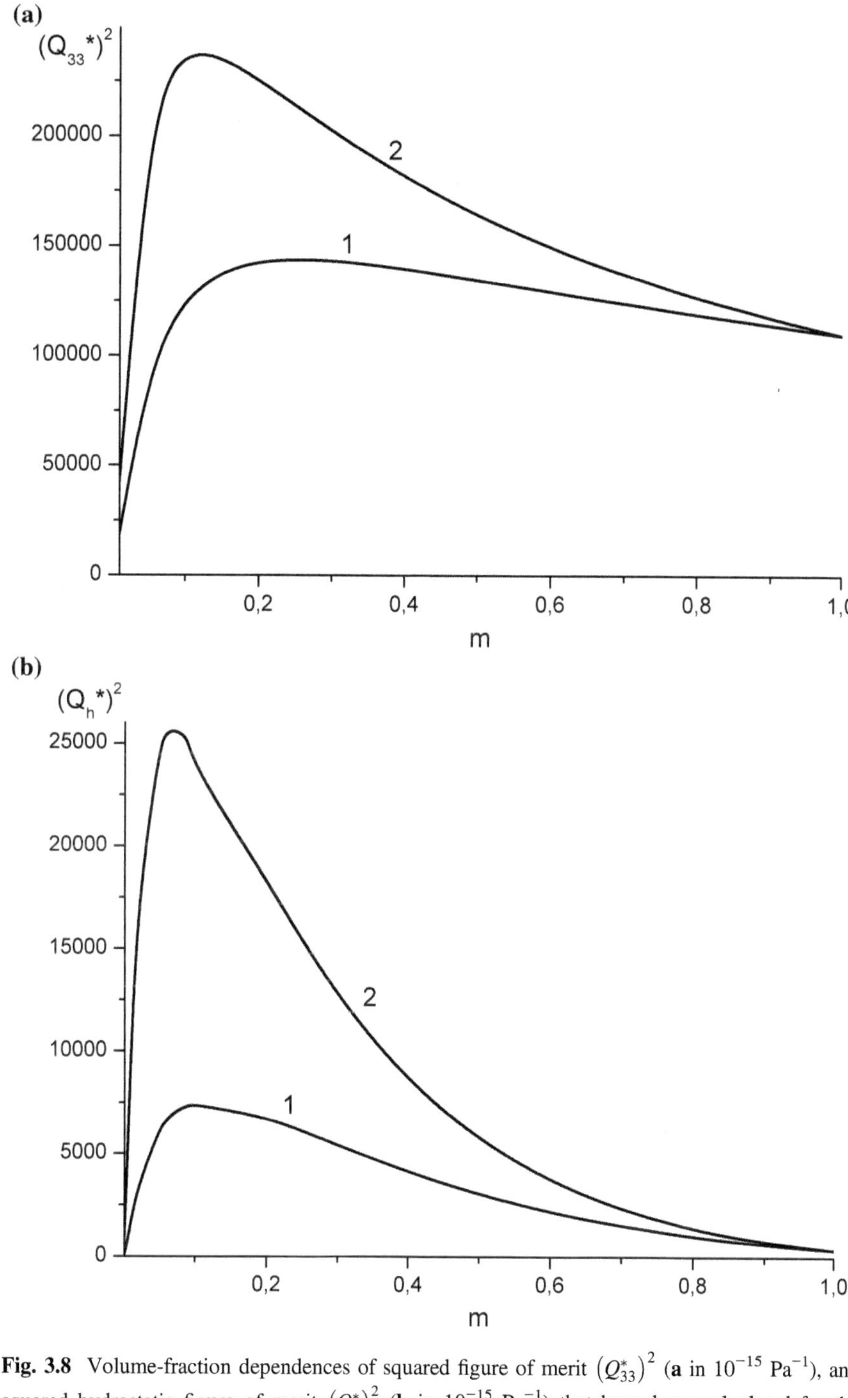

Fig. 3.8 Volume-fraction dependences of squared figure of merit $(Q_{33}^{*})^2$ (**a** in 10^{-15} Pa^{-1}), and squared hydrostatic figure of merit $(Q_h^{*})^2$ (**b** in 10^{-15} Pa^{-1}) that have been calculated for the following 1–3-type composites: PMN–0.33PT SC/araldite (curve 1) and PMN–0.33PT SC/porous araldite at the volume fraction $m_p = 0.3$ of spherical air pores (curve 2) (reprinted from Bezus et al. [42], with permission from IOP Publishing)

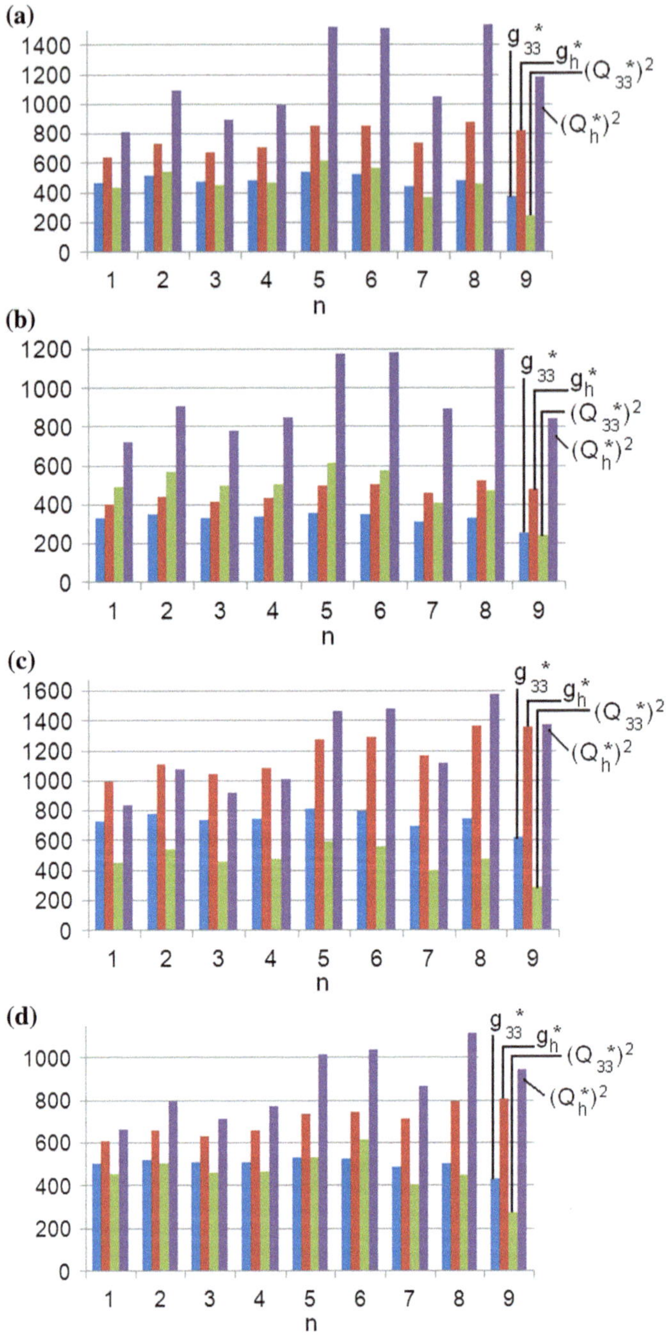

Fig. 3.9 Piezoelectric coefficients g_{33}^* and g_h^* (in mV m/N) and squared figures of merit $\left(Q_{33}^*\right)^2$ and $\left(Q_h^*\right)^2$ (in 10^{-13} Pa^{-1}) of 1–3-type KNN-TL SC/auxetic PE-n (graphs **a** and **c**) and KNN-T SC/auxetic PE-n (graphs **b** and **d**) composites at volume fractions of SC $m = 0.05$ (graphs **a** and **b**) and $m = 0.10$ (graphs **c** and **d**) (reprinted from paper by Topolov and Bowen [47], with permission from Elsevier)

performance is accounted for by the role of the elastic properties of the auxetic polymer component at its dielectric permittivity $\varepsilon_{33}^{(2)} \ll \varepsilon_{33}^{*\sigma}$. At $m \leq 0.1$, the dielectric permittivity of the 1–3-type composite obeys the condition $\varepsilon_{33}^{*\sigma} \ll \varepsilon_{33}^{(1)}$ that promotes large piezoelectric coefficients g_{33}^* and g_h^* and strongly influences the squared figures of merit $(Q_{33}^*)^2$ and $(Q_h^*)^2$ in accordance with (3.6).

3.4 0–3 Composites

Because of the system of isolated piezoelectric inclusions in a continuous matrix (Fig. 2.7), the 0–3 composite is characterised by moderate figures of merit. Despite the presence of the very prolate FC inclusions oriented along the poling axis OX_3, values of $\left(Q_{3j}^*\right)^2$ and $(Q_h^*)^2$ of the 0–3 composite based on the FC with a high piezoelectric activity are less than those of the related 1–3 composite [35]. The values of $\left(Q_{3j}^*\right)^2$ and $(Q_h^*)^2$ of the 0–3 PCR-7M FC/araldite composite (Table 3.9) are also less than $\left(Q_{3j}^*\right)^2$ and $(Q_h^*)^2$ of the 1–3-type PCR-7M FC/auxetic polyethylene composite; however they are comparable to $(Q_{3j}^*)^2$ and $(Q_h^*)^2$ of the 1–3 PZT-5 FC/araldite composite (Table 3.8). Despite the relatively large piezoelectric coefficient $g_{33}^{(1)}$ of the FC component, the 0–3 $(Pb_{0.75}Ca_{0.25})TiO_3$ FC/araldite composite is characterised by small values of $\left(Q_{3j}^*\right)^2$ and $(Q_h^*)^2$ in comparison to the 0–3 PCR-7M FC/araldite composite (Table 3.9). Decreasing $\left(Q_{3j}^*\right)^2$ and $(Q_h^*)^2$ from (3.6) may be due to the polymer (its elastic properties) or FC (its piezoelectric activity).

The 0–3 PMN–0.33PT SC/araldite composite with prolate inclusions exhibits high values of $\left(Q_{3j}^*\right)^2$ at $\rho = 0.1$ (Table 3.9), and increasing the aspect ratio ρ gives rise to an appreciable decrease of $\left(Q_{3j}^*\right)^2$ and $(Q_h^*)^2$. This effect observed in the presence of a SC component with high piezoelectric activity (see d_{3j} in Table 1.1) and is accounted for by the influence of the shape of inclusions on the piezoelectric properties of the 0–3 composite [7]. An increase of ρ leads to the effect of the depolarising field of the inclusion on the effective electromechanical properties of the 0–3 composite, and as a consequence, its piezoelectric activity decreases with increasing ρ.

The next example of the volume-fraction dependence of the squared figures of merit $\left(Q_{3j}^*\right)^2$ and $(Q_h^*)^2$ of composites based on the PMN–0.33PT SC is shown in

Table 3.9 Volume-fraction dependences of the piezoelectric coefficients d_{3j}^* (in pC/N), g_{3j}^* (in mV m/N) and squared figures of merit $\left(\left(Q_{3j}^*\right)^2\right.$ and $\left(Q_h^*\right)^2$ (in 10^{-12} Pa^{-1}), which have been calculated for the 0–3 composites with spheroidal piezoelectric inclusions

m	d_{31}^*	d_{33}^*	g_{31}^*	g_{33}^*	$(Q_{31}^*)^2$	$(Q_{33}^*)^2$	$(Q_h^*)^2$
PCR-7M FC/araldite composite, aspect ratio ρ = 0.01							
0	0	0	0	0	0	0	0
0.05	–98.5	268	–70.9	193	6.98	51.7	3.64
0.1	–151	405	–46.1	124	6.96	50.2	3.28
0.2	–210	544	–27.8	72.0	5.84	39.2	2.03
0.3	–244	614	–20.0	50.3	4.88	30.9	1.30
0.4	–267	657	–15.8	38.9	4.22	25.6	0.898
0.5	–286	686	–13.2	31.7	3.78	21.7	0.604
$(Pb_{0.75}Ca_{0.25})TiO_3$ FC/araldite composite, aspect ratio ρ = 0.01							
0	0	0	0	0	0	0	0
0.1	–6.58	21.5	–38.0	124	0.250	2.67	0.400
0.2	–6.54	24.6	–21.0	78.7	0.137	1.94	0.423
0.3	–5.90	25.8	–13.1	57.1	0.0773	1.47	0.433
0.4	–5.12	26.5	–8.66	44.8	0.0443	1.19	0.447
0.5	–4.30	26.9	–5.88	36.8	0.0253	0.990	0.458
PMN–0.33PT SC/araldite composite, aspect ratio ρ = 0.1 [48]							
0	0	0	0	0	0	0	0
0.05	–15.0	39.4	–94.7	249	1.42	9.81	0.542
0.1	–27.2	70.1	–81.9	211	2.23	14.8	0.741
0.2	–47.8	118	–59.1	146	2.82	17.2	0.614
0.3	–67.5	161	–46.3	110	3.13	17.7	0.452
0.4	–89.9	207	–38.5	88.6	3.46	18.3	0.316
0.5	–121	267	–33.6	73.8	4.07	19.7	0.165
PMN–0.33PT SC/araldite composite, aspect ratio ρ = 0.3 [48]							
0	0	0	0	0	0	0	0
0.05	–1.51	3.68	–26.2	63.9	0.0396	0.235	$7.59 \cdot 10^{-3}$
0.1	–3.04	7.25	–35.7	85.1	0.109	0.617	$1.06 \cdot 10^{-2}$
0.2	–6.01	13.8	–38.6	88.8	0.232	1.23	$2.06 \cdot 10^{-2}$
0.3	–9.12	20.4	–36.2	81.0	0.330	1.65	$1.86 \cdot 10^{-2}$
0.4	–13.0	28.3	–33.5	72.7	0.436	2.06	$1.31 \cdot 10^{-2}$
0.5	–20.5	43.1	–31.4	66.0	0.644	2.84	$6.72 \cdot 10^{-3}$
PMN–0.33PT SC/araldite composite, aspect ratio ρ = 0.5 [48]							
0	0	0	0	0	0	0	0
0.05	–0.570	1.31	–12.3	28.2	0.00701	0.0369	$7.20 \cdot 10^{-4}$
0.1	–1.19	2.68	–20.0	44.9	0.0238	0.120	$1.47 \cdot 10^{-3}$
0.2	–2.50	5.48	–27.1	59.3	0.0678	0.325	$2.45 \cdot 10^{-3}$

(continued)

Table 3.9 (continued)

m	d^*_{31}	d^*_{33}	g^*_{31}	g^*_{33}	$(Q^*_{31})^2$	$(Q^*_{33})^2$	$(Q^*_h)^2$
0.3	–3.99	8.57	–29.0	62.4	0.116	0.535	$2.60 \cdot 10^{-3}$
0.4	–5.97	12.7	–29.1	61.6	0.173	0.764	$2.58 . 10^{-3}$
0.5	–10.3	22.0	–29.8	61.4	0.307	1.35	$5.32 . 10^{-3}$

Notes

1. The effective piezoelectric coefficients and squared figures of merit of the PCR-7M FC/araldite composite and $(Pb_{0.75}Ca_{0.25})TiO$ FC/araldite composites were evaluated by means of the effective field method [7, 10]
2. The effective piezoelectric coefficients and squared figures of merit of the PMN–0.33PT SC/araldite composite were evaluated using the finite element method [10, 49]

Fig. 3.10. Changes in $(Q^*_h)^2$ are observed in relatively narrow ranges (see curves 2, 4 and 6 in Fig. 3.10) irrespective of the components, while $(Q^*_{33})^2$ can vary in wide ranges (see curves 1, 3 and 5 in Fig. 3.10). In our opinion, this behaviour is concerned with the important role of the matrix, its piezoelectric activity and anisotropy. It is also seen that the non-monotonic behaviour of $(Q^*_h)^2$ (see curve 2 in Fig. 3.10) is observed in the composite with the polymer matrix only. Replacing such a matrix with a FC or porous FC matrix leads to monotonic increasing $(Q^*_h)^2$ (see curves 4 and 6 in Fig. 3.10). It should be added that values of $(Q^*_h)^2$ from Fig. 3.10 exceed those related to the 0–3 $PbTiO_3$-based and 1–3–0 PZT rod composites [14].

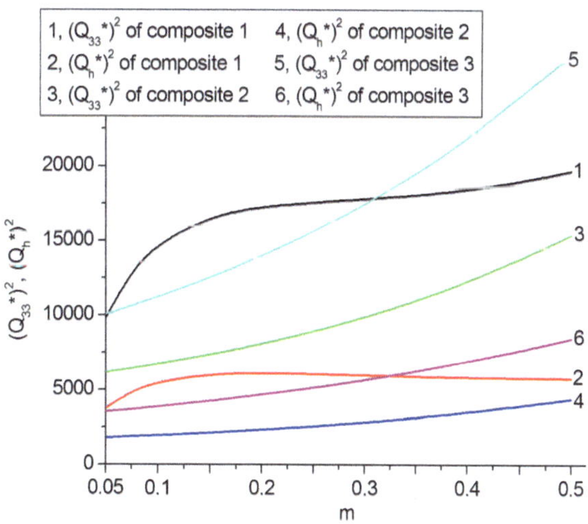

Fig. 3.10 Squared figures of merit $(Q^*_{33})^2$ and $(Q^*_h)^2$ (in 10^{-15} Pa^{-1}) which have been calculated using the finite element method data on piezoelectric coefficients d^*_{3j} and g^*_{3j} of the following 0–3-type composites with spheroidal SC inclusions at their aspect ratio $\rho = 0.1$: PMN–0.33PT SC/araldite (termed 'composite 1'), PMN–0.33PT SC/PMN–0.35PT FC (termed 'composite 2') and PMN–0.33PT SC/porous PMN–0.35PT FC at the volume fraction $m_p = 0.3$ of spherical air pores (termed 'composite 3') (reprinted from paper by Topolov et al. [49], with permission from Taylor and Francis)

3.5 New Effects and Improved Figures of Merit in 1–3-Type Composites Based on Relaxor-Ferroelectric Single Crystals

3.5.1 New Orientation Effect

In Sect. 3.5.1 the influence of the mutual orientation of the poling directions of relaxor-ferroelectric SC and FC components on the electromechanical properties and related parameters of the 1–3-type composites is considered. These composites comprise two ferroelectric components with different polarisation directions and appreciable differences in their electromechanical properties, which lead to a *new orientation effect* [44] and improve the effective parameters. This additional improvement is of an independent interest because of complex inter-relationships in the fundamental triangle [7] of 'composition—structure—properties'.

It is assumed that the composite consists of long SC rods embedded in a heterogeneous matrix (Fig. 3.11a). The SC rods are in the form of the rectangular parallelepiped with a square base and square arrangement in the (X_1OX_2) plane. The main crystallographic axes X, Y and Z of each SC rod with the spontaneous polarisation $\boldsymbol{P}_s^{(1)}$ are oriented on conditions that X || OX_1, Y || OX_2 and Z || $\boldsymbol{P}_s^{(1)}$ || OX_3. The FC component is used as an inclusion in the polymer matrix that surrounds the system of the aligned SC rods. The shape of each FC inclusion is spheroidal and obeys the equation $(x_1'/a_1)^2 + (x_2'/a_2)^2 + (x_3'/a_3)^2 = 1$ relative to the axes of the rectangular co-ordinate system $(X_1'X_2'X_3')$ rotated by an angle α with respect to $(X_1X_2X_3)$ (inset 1 in Fig. 3.11a). The semi-axes of each FC inclusion are $a_1 = a_2$ and a_3, the aspect ratio of each FC inclusion is $\rho_i = a_1/a_3$, and centres of the inclusions (Fig. 3.11b) occupy sites of a simple tetragonal lattice with unit-cell vectors parallel to the OX_k' axes. We assume that $0 < \rho_i < 1$, and the presence of prolate inclusions facilitates poling of the FC/polymer matrix due to a weaker depolarisation effect in the electric field. The remanent polarisation vector of each FC inclusion is $\boldsymbol{P}_r^{(2)} \uparrow\uparrow OX_3'$, and OX_3' is the poling axis of the matrix (inset 2 in Fig. 3.11a) that represents a composite with 0–3 connectivity in terms of work [7, 13]. The three-component composite shown in Fig. 3.11a is described by 1–0–3 connectivity. It is assumed that the electrodes applied to the composite are perpendicular to the OX_3 axis. By assuming that the linear sizes of the FC inclusions in the 0–3 matrix are significantly smaller than the length of the side of the square being intersected the rod in the (X_1OX_2) plane, we evaluate the effective properties of the 1–0–3 composite in two stages.

At the first stage of averaging, we take into account the electromechanical interaction between the FC inclusions in the 0–3 matrix and evaluate its effective properties by means of the effective field method [7, 10]. An alternative way to find the effective properties of the 0–3 matrix is concerned with the use of the finite element method. COMSOL [44] was applied to obtain the volume-fraction dependence of the effective electromechanical properties of the 0–3 composite within the finite element method framework. The matrix of the effective

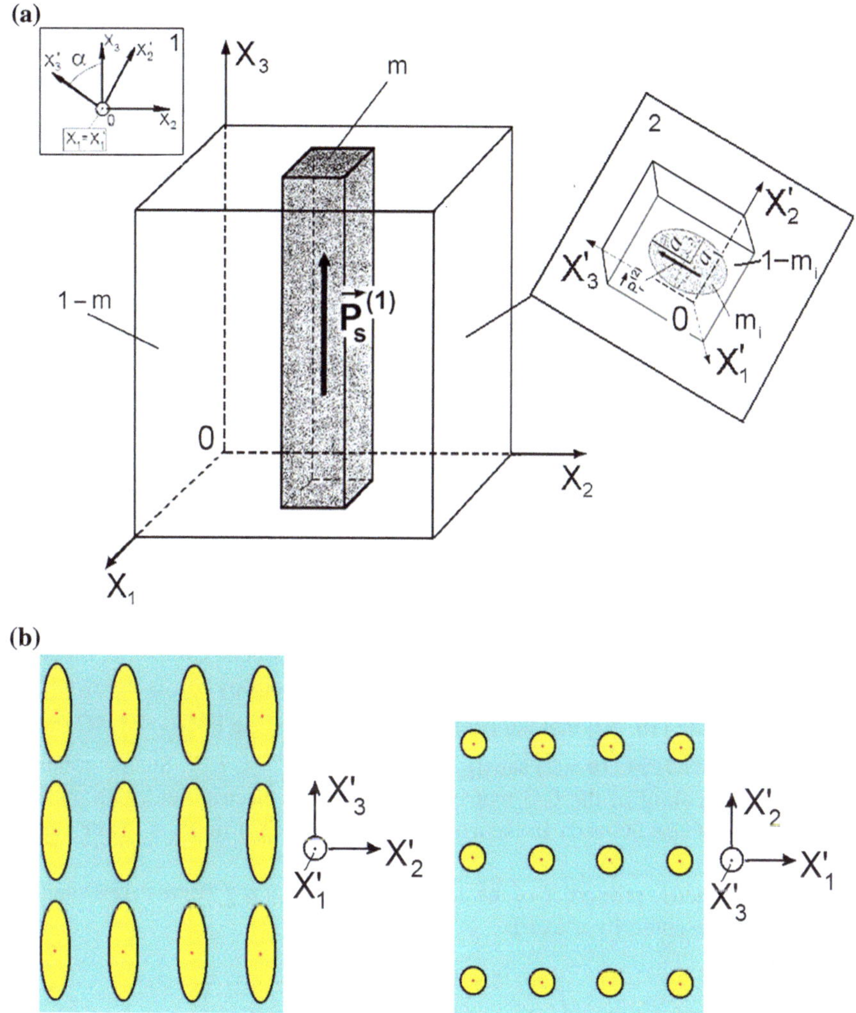

Fig. 3.11 Schematic of the 1–0–3 SC/FC/polymer composite (**a**), and examples of the regular arrangement of spheroidal FC inclusions in the 0–3 matrix along the co-ordinate axes OX_k'. m and $1 - m$ are volume fractions of the SC and surrounding 0–3 matrix, respectively. Rotation of co-ordinate axes $(X_1'X_2'X_3') \rightarrow (X_1X_2X_3)$ is shown in *inset 1*, and the 0–3 matrix is shown in *inset 2*. m_i and $1 - m_i$ are volume fractions of FC and polymer, respectively, in the 0–3 matrix (reprinted from paper by Topolov et al. [44], with permission from Elsevier)

electromechanical properties $\left\| C_{0-3}^{*} \right\|$ determined using either the effective field method of finite element method represents a function of the volume fraction of FC m_i in the 0–3 composite and the aspect ratio of the FC inclusion ρ_i therein. Taking into account the rotation $(X_1'X_2'X_3') \rightarrow (X_1X_2X_3)$ (see inset 1 in Fig. 3.11a), we find the matrix of the effective electromechanical properties of the 0–3 composite $\|C^*_{0-3}\| = \|C^*_{0-3}(m_i, \rho_i, \alpha)\|$ in the co-ordinate system $(X_1X_2X_3)$.

At the second stage of averaging, the effective electromechanical properties of the 1–0–3 composite with planar interfaces (Fig. 3.11a) are evaluated using the matrix method [7, 10]. The electromechanical properties of the SC rod and 0–3 composite matrix are averaged in the OX_1 and OX_2 directions, in which the periodic structure of the composite is observed. Hereby we take into account electromechanical interactions in a system of 'piezo-active rods—piezo-active matrix'. The effective electromechanical properties of the 1–0–3 composite in the co-ordinate system $(X_1X_2X_3)$ are given by

$$\|C^*\| = \|C^*(m, m_i, \alpha)\| = \begin{pmatrix} \|s^{*E}\| & \|d^*\|^t \\ \|d^*\| & \|\varepsilon^{*\sigma}\| \end{pmatrix}$$

Among the components of interest for the 1–0–3 composite shown in Fig. 3.11a, we choose the [001]-poled domain-engineered PMN–0.33PT SC, poled $(Pb_{1-x}Ca_x)$ TiO_3 FC and araldite. Their properties are shown in Tables 1.1, 2.7 and 3.10. The PMN–0.33PT SC exhibits a high piezoelectric activity and moderate piezoelectric anisotropy (see data from Table 1.1). The $(Pb_{1-x}Ca_x)TiO_3$ FC with $0.20 \le x \le 0.25$ is the second component due to its contrasting properties, compared to PMN–0. 33PT SC. As follows from Table 3.10, the FC component exhibits a moderate piezoelectric activity with a large piezoelectric anisotropy. It is known from experimental data [16, 50] that the coercive fields $E_c^{(n)}$ of the PMN–xPT SC ($n = 1$) and $(Pb_{1-x}Ca_x)TiO_3$ FC ($n = 2$) satisfy the condition $E_c^{(1)} \ll E_c^{(2)}$. Such a condition enables initial poling of the 0–3 matrix under a strong electric field with a subsequent poling of the SC rods in the composite (Fig. 3.11a) under a lower electric field.

The piezoelectric properties of the 1–0–3 composite at $\alpha \ne 0°$, $\alpha \ne 180°$ and $0 < m < 1$ are represented by a matrix

$$\|p*\| = \begin{pmatrix} 0 & 0 & 0 & 0 & p^*_{15} & p^*_{16} \\ p^*_{21} & p^*_{22} & p^*_{23} & p^*_{24} & 0 & 0 \\ p^*_{31} & p^*_{32} & p^*_{33} & p^*_{34} & 0 & 0 \end{pmatrix}$$

Table 3.10 Room-temperature elastic compliances s^E_{ab} (in 10^{-12} Pa^{-1}), piezoelectric coefficients d_{fl} (in pC/N) and relative dielectric permittivities $\varepsilon^\sigma_{rr}/\varepsilon_0$ of the poled FC component of the 0–3 matrix

Component	s^E_{11}	s^E_{12}	s^E_{13}	s^E_{33}	s^E_{44}	s_{66}	d_{31}	d_{33}	d_{15}	$\varepsilon^\sigma_{11}/\varepsilon_0$	$\varepsilon^\sigma_{33}/\varepsilon_0$
$(Pb_{0.80}Ca_{0.20})\cdot$ $\cdot TiO_3$ FC [41]	6.04	−1.24	−1.25	6.21	14.7	14.6	−1.33	24.6	26.1	131	135
$(Pb_{0.75}Ca_{0.25})\cdot$ $\cdot TiO_3$ FC [41]	6.00	−1.30	−1.30	6.18	14.8	14.6	−0.364	28.0	28.9	158	163

where $p = d, e, g$, or h. Taking into account the rotation mode (see inset 1 in Fig. 3.11a) and symmetry of the components, we find that the effective properties of the 1–3–0 composite Π^* obey the condition $\Pi^*(m, m_i, \rho_i, \alpha) = \Pi^*(m, m_i, \rho_i, 360° - \alpha)$ [44]. Hereafter we consider examples of the orientation (α) and volume-fraction (m or m_i) dependences of its effective parameters at ρ_i = const.

The orientation dependence of the hydrostatic parameters of the 1–0–3 composite suggests that max g_h^* and max $\left[\left(Q_h^*\right)^2\right]$ are achieved at $\alpha \approx 90°$ and $m_i = 0.12$ (Fig. 3.12). The largest value of g_h^* at $0.1 \le \rho_i \le 0.5$ is related to $m_i \approx 0.12$ and $0.01 < m < 0.03$, and in this m range a local max g_h^* is observed at various values of m_i, ρ_i and α (Fig. 3.12). Our evaluations based on the effective field method (for the 0–3 matrix) and matrix method (for the 1–3–0 composite) lead to absolute max $d_h^* = 305$ pC/N at $m = 0.532$, $m_i = 0.12$, $\rho_i = 0.1$, and $\alpha = 90°$ [44].

Using the matrix method, we find for the related 1–3 PMN–0.33PT SC/araldite composite (i.e., in a case of $m_i = 0$ in Fig. 3.11a), that absolute max $g_h^* = 158$ mV m/N, max $\left[\left(Q_h^*\right)^2\right] = 8.27 \cdot 10^{-12}\ \mathrm{Pa}^{-1}$ and max $d_h^* = 274$ pC/N are achieved at $m = 0.016$, 0.103 and 0.509, respectively. At $m = 0.05$ for the 1–3 PMN–0.33PT SC/araldite composite, we find $g_h^* = 115$ mV m/N, and this value is considerably smaller than g_h^* near the maxima in Fig. 3.12a. The large values of $\left(Q_h^*\right)^2$ (Fig. 3.12b) and d_h^* in the 1–0–3 composite are due to the presence of the 0–3 matrix that comprises of a FC component with piezoelectric coefficients $d_{3f}^{(2)}$ that obey the condition [41] $d_{33}^{(2)} / \left|d_{31}^{(2)}\right| \gg 1$. At the rotation angle $\alpha = 90°$, the remanent polarisation vector $\boldsymbol{P}_r^{(2)}$ of each FC inclusion (see inset 2 in Fig. 3.11a) is parallel to OX_2, and this $\boldsymbol{P}_r^{(2)}$ orientation leads to a decrease in $\left|d_{32}^*\right|$ with minor changes in d_{31}^* and d_{33}^* (or g_{31}^* and g_{33}^*, respectively) as a result of the weak lateral piezoelectric effect in the 0–3 FC/polymer matrix. As a consequence of the reduced $\left|d_{32}^*\right|$, we observe an increase in both d_h^* and g_h^* of the 1–3–0 composite.

We note that the elastic anisotropy of the 0–3 matrix with highly prolate FC inclusions is an additional factor in increasing the hydrostatic parameters of the 1–0–3 composite. For instance, ratios of the elastic compliances of a 0–3 $(Pb_{0.75}Ca_{0.25})TiO_3$ FC/araldite composite with spheroidal inclusions (see Fig. 3.11b) are $s_{11,0-3}^E / s_{12,0-3}^E = -2.21$, $s_{11,0-3}^E / s_{13,0-3}^E = -6.52$ and $s_{11,0-3}^E / s_{33,0-3}^E = 1.95$ at $\rho_i = 0.1$ and $m_i = 0.10$. At $\rho_i = 0.3$ and $m_i = 0.10$, for the same composite we find $s_{11,0-3}^E / s_{12,0-3}^E = -2.52$, $s_{11,0-3}^E / s_{13,0-3}^E = -3.51$ and $s_{11,0-3}^E / s_{33,0-3}^E = 1.25$. It is seen that a significant decrease of $\left|s_{11,0-3}^E / s_{13,0-3}^E\right|$ and $\left|s_{11,0-3}^E / s_{33,0-3}^E\right|$ takes place when the piezoelectric activity decreases. This orientation effect in the 1–0–3 composite favours an increase in $\left(Q_h^*\right)^2$ and d_h^* near its maxima by approximately 29 and 11 %, respectively [44], in comparison to the aforementioned 1–3 PMN–0.33PT SC/araldite composite. The studied 1–0–3

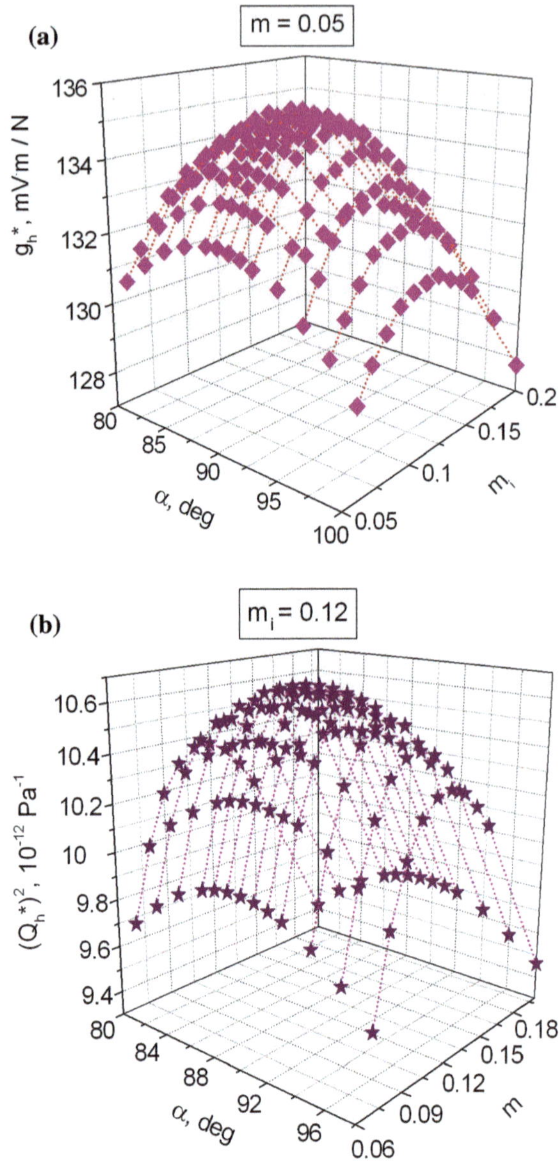

Fig. 3.12 Examples of the orientation dependence of the hydrostatic parameters of the 1–0–3 PMN–0.33PT SC/$(Pb_{0.75}Ca_{0.25})TiO_3$ FC/araldite composite at $\rho_i = 0.1$: (**a**) near local $\max g_h^*(0.05, m_i, 0.1, \alpha)$ and (**b**) near absolute $\max\left\{\left[Q_h^*(m, 0.12, 0.1, \alpha)\right]^2\right\}$. Electromechanical properties of the 0–3 matrix at the first stage of averaging were determined by means of the effective field method (reprinted from paper by Topolov et al. [44], with permission from Elsevier)

composite is also attractive due to large values of local maxima of d_h^*, g_h^* and $\left(Q_h^*\right)^2$ at volume fractions of FC inclusions m_i = const, especially at $m_i < 0.15$.

Replacing the $(Pb_{0.75}Ca_{0.25})TiO_3$ FC with $(Pb_{1-x}Ca_x)TiO_3$ at $x = 0.20–0.24$ leads to changes in the hydrostatic parameters of the 1–0–3 composite by 1–3 % at fixed values of m, m_i and ρ_i. The $(Pb_{1-x}Ca_x)TiO_3$ FCs at $x = 0.20–0.24$ are characterised by the large piezoelectric anisotropy: e.g., $d_{33}^{(2)}/\left|d_{31}^{(2)}\right|$ = 18.7, 30.3 and 39.8 at

$x = 0.20$, 0.23 and 0.24, respectively [41]. These FCs exhibit an appreciable elastic anisotropy that is typical of modified $PbTiO_3$ FCs [50]. Our results suggest that the large piezoelectric anisotropy of the $(Pb_{1-x}Ca_x)TiO_3$ FC leads to a large piezoelectric anisotropy of the 0–3 composite and favours the orientation effect in the 1–0–3 composite and improving their hydrostatic parameters in comparison to the 1–3 composite.

3.5.2 New Aspect-Ratio Effect

In Sect. 3.5.2 we discuss an influence of the aspect ratio of FC inclusions ρ_i in the 0–3 matrix (see inset 2 in Fig. 3.11a) on the piezoelectric performance and hydrostatic parameters of the 1–0–3 composite. In our study, the [001]-poled domain-engineered PMN–0.33PT SC is regarded as the main piezoelectric component (see the full set of its electromechanical constants in Table 1.1), the modified $PbTiO_3$ FC is the main component in the 0–3 matrix (see the full set of electromechanical constants of the modified $PbTiO_3$ (I) FC in Table 1.2), and the polymer component is either araldite or polyurethane (their properties are shown in Table 2.7). It is obvious that the SC and FC components are selected due to their contrasting electromechanical properties.

An important example of the dependence of the piezoelectric coefficients g_{33}^* and g_h^* and squared figures of merit $\left(Q_{3j}^*\right)^2$ on the aspect ratio ρ_i is shown in Fig. 3.13. Distinct changes in the effective parameters of the 1–0–3 composite are observed [45] at

$$0.01 < \rho_i < 2. \tag{3.12}$$

The shape of the FC inclusion in the aspect-ratio range (3.12) changes from highly prolate (ρ_i << 1) to oblate (ρ_i > 1). These changes in the microgeometry of the 0–3 matrix give rise to significant changes in its elastic properties and ratios of elastic compliances of the matrix $s_{1b}^{(m),E}/s_{kl}^{(m),E}$ which have a strong influence on the piezoelectric properties and their anisotropy in the 1–0–3 composite as a whole. We add that some examples of the influence of the elastic anisotropy of the 0–3 matrix on the performance of the 1–0–3 composite were considered in Sect. 3.5.1.

Now we compare some effective parameters which are calculated using different methods [45]. In the first case we calculate the effective properties of the 0–3 matrix by means of the finite element method and then the effective properties of the 1–0–3 composite as a whole by means of the matrix method. In the second case we apply the effective field method to evaluate the effective properties of the 0–3 matrix and the matrix method to find the effective properties of the 1–0–3 composite. Example results are given for comparison in Table 3.11. As follows from Table 3.11, the effective parameters concerned with the longitudinal [g_{33}^* and $\left(Q_{33}^*\right)^2$], lateral

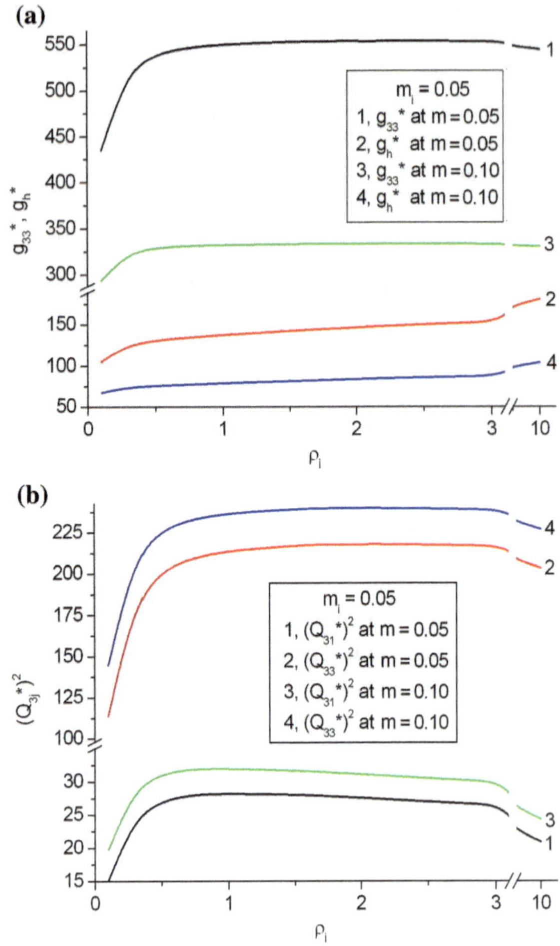

Fig. 3.13 Aspect-ratio dependence of the piezoelectric coefficient g_{33}^* and hydrostatic piezoelectric coefficient g_h^* (**a** in mV m/N), and squared figures of merit $\left(Q_{3j}^*\right)^2$ (**b** in 10^{-12} Pa^{-1}) of the 1–0–3 PMN–0.33PT SC/modified $PbTiO_3$ FC/polyurethane composite at volume fractions m_i = const (FC inclusions in the 0–3 matrix) and m = const (SC rods in the composite), see the schematic in Fig. 3.11a. At the first stage of averaging, electromechanical properties of the 0–3 matrix were evaluated by means oft he effective field method (reprinted from paper by Topolov et al. [45], with permission from Elsevier)

$\left[\left(Q_{31}^*\right)^2\right]$ and hydrostatic $\left[\left(Q_h^*\right)^2\right]$ piezoelectric effects of the 1–0–3 composite are in agreement. The largest relative error (δ = 10–11 %) is related to squared figures of merit $\left(Q_{3j}^*\right)^2$ and $\left(Q_h^*\right)^2$. These parameters combine the piezoelectric activity and sensitivity. Maxima of $\left(Q_{3j}^*\right)^2$ and $\left(Q_h^*\right)^2$ are observed at relatively small volume fractions m of SC (approximately 0.1) due to a non-monotonic behaviour of g_{3j}^* and g_h^*, respectively. The location of max g_{33}^*, min g_{31}^* and max g_h^* at $m \ll 1$ (as a rule, less than 0.05) strongly depends on the elastic and dielectric properties of the 0–3 matrix. It is seen that differences between the parameters calculated for the 1–0–3 composite using different methods are located near maxima of $\left(Q_{3j}^*\right)^2$ and $\left(Q_h^*\right)^2$, i.e., at m = 0.05 or 0.10 (Table 3.11).

Table 3.11 Comparison of effective parameters calculated for the 1–0–3 PMN–0.33PT SC/modified $PbTiO_3$ FC/polymer composite using either the finite element method (Π^*_{FEM}) or effective field method (Π^*_{EFM}) to evaluate properties of the 0–3 matrix

Evaluation method for 0–3 matrix properties	ρ_i	m_i	m	g^*_{33}, mV m/N	$(Q^*_{33})^2$, 10^{-12} Pa^{-1}	$(Q^*_{31})^2$, 10^{-12} Pa^{-1}	$(Q^*_h)^2$, 10^{-12} Pa^{-1}
1–0–3 PMN–0.33PT SC/modified $PbTiO_3$ FC/araldite composite							
Finite element method	1.5	0.05	0.05	433	96.6	12.3	7.03
	2.0	0.05	0.05	433	96.4	12.1	7.30
	2.5	0.05	0.05	432	95.7	11.8	7.57
	1.5	0.10	0.05	406	81.3	9.84	6.75
	2.0	0.10	0.05	405	80.6	9.45	7,22
	2.5	0.10	0.05	403	79.2	8.98	7.65
	1.5	0.05	0.10	291	126	16.6	8.27
	2.0	0.05	0.10	291	126	16.3	8.58
	2.5	0.05	0.10	291	125	15.9	8.90
	1.5	0.10	0.10	281	110	13.8	8,21
	2.0	0.10	0.10	281	109	13.3	8.77
	2.5	0.10	0.10	280	108	12.7	9.30
Effective field method	1.5	0.05	0.05	438	98.8	12.4	7.39
	2.0	0.05	0.05	439	99.2	12.2	7.77
	2.5	0.05	0.05	439	99.1	12.0	8.06
	1.5	0.10	0.05	418	86.8	10.4	7.33
	2.0	0.10	0.05	420	87.5	10.2	7.97
	2.5	0.10	0.05	420	87.5	9.88	8.50
	1.5	0.05	0.10	293	128	16.6	8.61
	2.0	0.05	0.10	293	128	16.4	9.04
	2.5	0.05	0.10	293	128	16.1	9.37
	1.5	0.10	0.10	285	115	14.4	8.47
	2.0	0.10	0.10	286	116	14.0	9.48
	2.5	0.10	0.10	286	116	13.6	10.1
1–0–3 PMN–0.33PT SC/modified $PbTiO_3$ FC/polyurethane composite							
Finite element method	1.5	0.05	0.05	549	212	27.7	13.8
	2.0	0.05	0.05	549	212	27.2	14.4
	2.5	0.05	0.05	548	210	26.5	15.0
	1.5	0.10	0.05	525	182	22.6	13.7
	2.0	0.10	0.05	524	180	21.6	14.8
	2.5	0.10	0.05	522	177	20.4	15.8
	1.5	0.05	0.10	332	235	31.5	13.8
	2.0	0.05	0.10	332	235	30.9	14.4
	2.5	0.05	0.10	332	233	30.2	15.0
	1.5	0.10	0.10	325	209	26.7	14.2
	2.0	0.10	0.10	324	207	25.6	15.2
	2.5	0.10	0.10	324	204	24.4	16.3

(continued)

Table 3.11 (continued)

Evaluation method for 0–3 matrix properties	ρ_i	m_i	m	g^*_{33}, mV m/N	$(Q^*_{33})^2$, 10^{-12} Pa^{-1}	$(Q^*_{31})^2$, 10^{-12} Pa^{-1}	$(Q^*_h)^2$, 10^{-12} Pa^{-1}
Effective field method	1.5	0.05	0.05	553	217	28.1	14.4
	2.0	0.05	0.05	553	218	27.5	15.3
	2.5	0.05	0.05	553	218	27.0	16.1
	1.5	0.10	0.05	531	194	23.7	14.8
	2.0	0.10	0.05	536	195	23.0	16.3
	2.5	0.10	0.05	536	195	22.2	17.5
	1.5	0.05	0.10	333	240	31.8	14.4
	2.0	0.05	0.10	334	240	31.2	15.2
	2.5	0.05	0.10	334	240	30.6	16.0
	1.5	0.10	0.10	328	219	27.7	15.1
	2.0	0.10	0.10	328	220	26.8	16.6
	2.5	0.10	0.10	328	220	26.0	17.8

Replacing araldite with a more compliant polyurethane (their elastic properties are shown in Table 2.7) in the 0–3 matrix leads to an increase of all of the aforementioned effective parameters of the 1–0–3 composite (see Table 3.11). Such an effect is due to the more pronounced longitudinal and hydrostatic piezoelectric effects in a 1–3-type composite [7] that contains a more compliant matrix which allows a larger free deformation of the embedded piezoelectric rods. The use of a more compliant matrix has a significant influence on the piezoelectric response of the 1–0–3 composite along different co-ordinate axes.

Table 3.11 suggests that the relation

$$\left(Q^*_{33}\right)^2/\left(Q^*_{31}\right)^2 \approx 8-9 \tag{3.13}$$

holds at various values of m, m_i and ρ_i. A relatively large $\left(Q^*_{33}\right)^2/\left(Q^*_{31}\right)^2$ ratio is achieved due to the effect of the elastic properties of the 0–3 matrix on the lateral piezoelectric response. The oblate FC inclusions in this matrix can weaken the piezoelectric coefficient d^*_{31} of the 1–0–3 composite due to the aforementioned elastic anisotropy of the 0–3 matrix (see Sect. 3.5.1).

Of particular interest is a case of a piezo-passive 0–3 matrix to examine the influence of the FC inclusion elastic properties on the composite performance. Now it is assumed that the FC inclusions in a polymer medium (inset 2 in Fig. 3.11a) have not been poled and, therefore, remain piezo-passive and isotropic. By varying the aspect ratio ρ_i of these inclusions it is possible to observe changes in the effective parameters of the 1–0–3 composite (Table 3.12), however these parameters become larger than those in the case of the poled (piezo-active) 0–3 FC/polymer matrix. This is a result of a decrease in the dielectric permittivity of the piezo-passive 0–3 matrix that leads to an appreciable decrease of $\varepsilon^{*\sigma}_{33}$ of the 1–0–3

Table 3.12 Effective parameters calculated for the 1–0–3 PMN–0.33PT SC/modified $PbTiO_3$ FC/polyurethane composite in a case of a piezo-passive 0–3 matrix[a] (reprinted from paper by Topolov et al. [45], with permission from Elsevier)

p_i	m_i	m	g_{33}^*, mV m/N	$(Q_{33}^*)^2$, 10^{-12} Pa^{-1}	$(Q_{31}^*)^2$, 10^{-12} Pa^{-1}	d_h^*, pC/N	g_h^*, mV m/N	$(Q_h^*)^2$, 10^{-12} Pa^{-1}
1.5	0.05	0.05	553	218	28.1	102	143	14.5
2.0	0.05	0.05	554	218	27.5	105	147	15.4
2.5	0.05	0.05	554	218	27.0	107	150	16.1
1.5	0.10	0.05	535	194	23.7	101	148	14.9
2.0	0.10	0.05	536	185	23.0	105	155	16.3
2.5	0.10	0.05	536	195	22.2	109	161	17.6
1.5	0.05	0.10	333	240	31.7	177	81.9	14.5
2.0	0.05	0.10	334	240	31.2	182	84.1	15.3
2.5	0.05	0.10	334	240	30.6	186	86.1	16.0
1.5	0.10	0.10	328	220	27.7	176	86.2	15.2
2.0	0.10	0.10	328	221	26.8	195	90.2	16.6
2.5	0.10	0.10	328	220	25.9	191	93.5	17.9

[a]Electromechanical properties of the 0–3 matrix were evaluated by means of the effective field method

composite at relatively small SC volume fractions. Results given in Table 3.12 suggest that the role of the elastic anisotropy in achieving high piezoelectric performance for the 1–0–3 composite remains dominant irrespective of the piezoelectric activity of its 0–3 matrix. Thus, during the manufacture of the 1–0–3 composite shown in Fig. 3.11a, there is no need to pole the 0–3 matrix under a relatively high electric field, as applied, for instance, to the $PbTiO_3$-type FC composite systems [50]. An incompleteness of the poling of the 0–3 matrix is avoided by this way, and a smaller dielectric permittivity $\varepsilon_{33}^{*\sigma}$ in the 1–0–3 composite favours its higher piezoelectric sensitivity.

The performance of the 1–0–3 composite is compared to data [51] on a 1–0–3 FC/FC/polyurethane composite where the highly piezo-active PCR-7M FC (a PZT-type composition, see data in Table 1.2) is the main component. Parameters that characterise the piezoelectric sensitivity of the 1–0–3 PCR-7M FC/PCR-7M FC/polyurethane composite are $g_{33}^* \approx 400$ mV m/N, $g_h^* \approx 200$ mV m/N, and the hydrostatic piezoelectric activity is characterized by $d_h^* \approx 350$ pC/N [51]. These effective parameters are comparable to those related to the 1–0–3 relaxor-ferroelectric SC/FC/polyurethane composite [45] that consists of contrasting components. The maximum value of g_{33}^* determined for a 1–3 PMN–0.30PT/epoxy composite [19] is 440 mV m/N (at the volume fraction of SC $m = 0.018$) and is comparable to values of g_{33}^* from Tables 3.11 and 3.12. A 1–3 composite based on PMN–PT SC is characterised by $d_h^* = 111$ pC/N, $g_h^* = 37$ mV m/N and $(Q_h^*)^2 = 4.12 \cdot 10^{-12}$ Pa^{-1} [52], and these parameters are smaller than those given in Tables 3.11 and 3.12. It should be added that the $(Q_h^*)^2$ values shown in Tables 3.11

and 3.12 are about 1.5–3 times more than $\left(Q_h^*\right)^2$ related to conventional 1–3 FC/polymer composites [7, 8, 13–15].

The next example of high-performance composite systems is related to the 1–0–3 composite based on the [001]-poled PZN–0.08PT SC [46]. This SC if of specific interest because of a high piezoelectric activity (for example, the longitudinal piezoelectric coefficient $d_{33}^{(1)}$ = 2890 pC/N, see Table 1.1) and small negative hydrostatic piezoelectric coefficients $d_h^{(1)}$ = –20 pC/N and $g_h^{(1)}$ = –0.293 mV m/N. It is now assumed that the 0–3 matrix surrounding the SC rods (Fig. 3.11a) contains modified $PbTiO_3$ FC inclusions in a polymer medium. The electromechanical properties of the 0–3 matrix wherein a regular arrangement of the spheroidal FC inclusions is assumed (Fig. 3.11b), are evaluated using either the finite element method or effective field method, and the subsequent averaging procedure for the system 'rods—matrix' is carried out using the matrix method. Results of our calculations of squared figures of merit and related parameters of the PZN–0.08PT-based composite are shown in Tables 3.13 and 3.14.

Data from Tables 3.13 and 3.14 show the active role of the matrix subsystem in achieving large piezoelectric coefficients and figures of merit of the 1–0–3 composite. Even at relatively small volume fractions of the FC inclusions m_i in the 0–3 matrix, changes in the aspect ratio ρ_i lead to changes in the aforementioned parameters of the 1–0–3 composite. We see that these changes take place irrespective of the piezoelectric activity of the 0–3 matrix. In general, a good correlation between the effective parameters calculated using different methods (see Tables 3.13 and 3.14) is observed. The smaller dielectric permittivity $\varepsilon_{33}^{*\sigma}$ (due to the lack of the piezoelectric contribution into the dielectric permittivity of the 0–3 matrix $\varepsilon_{33}^{(m)}$) favours a higher piezoelectric sensitivity of the 1–0–3 composite in accordance with the relation [11, 12] $g_{3j}^* = d_{3j}^*/\varepsilon_{33}^{*\sigma}$ (j = 1, 2 and 3). This leads to an increase in $\left(Q_{3j}^*\right)^2$ and $\left(Q_h^*\right)^2$ (Table 3.14) in comparison to those (Table 3.13) related to the composite with a piezo-active 0–3 matrix. Moreover, the elastic properties of the polymer component and, therefore, the 0–3 matrix strongly influences the studied parameters of the 1–0–3 composite: we see the larger values of $\left(Q_{3j}^*\right)^2$ and $\left(Q_h^*\right)^2$ in the presence of the softer, polyurethane, component.

We add for comparison, that at m_i = 0 (i.e., for the 1–3 PZN–0.08PT SC/polyurethane composite), absolute maxima are achieved as follows [46]: max d_h^* = 254 pC/N (at the volume fraction of SC m = 0.341), max g_h^* = 145 mV m/N (at m = 0.012) and max $\left[\left(Q_h^*\right)^2\right]$ = 9.53 · 10^{-12} Pa^{-12} (at m = 0.088). These values are considerably less than those given in Tables 3.13 and 3.14. Our comparison of data from Tables 3.12 and 3.14 suggests that in the 1–0–3 PZN–0.08PT SC/modified $PbTiO_3$ FC/polyurethane composite, larger values of its squared figures of merit

Table 3.13 Effective parameters calculated for the 1–0–3 PZN–0.08PT SC/modified $PbTiO_3$ FC/polymer composite in a case of a piezo-active 0–3 matrix[a]

ρ_i	m_i	m	g_{33}^*, mV m/N	$(Q_{33}^*)^2$, 10^{-12} Pa^{-1}	$(Q_{31}^*)^2$, 10^{-12} Pa^{-1}	d_h^*, pC/N	g_h^*, mV m/N	$(Q_h^*)^2$, 10^{-12} Pa^{-1}
1–0–3 PZN–0.08PT SC/modified $PbTiO_3$ FC/araldite composite								
1.5	0.05	0.05	287	56.1	5.68	71.2	104	7.43
2.0	0.05	0.05	286	56.0	5.68	70.9	104	7.37
2.5	0.05	0.05	285	55.5	5.69	70.0	103	7.18
1.5	0.10	0.05	264	46.5	5.75	52.2	78.1	4.08
2.0	0.10	0.05	263	46.0	5.77	51.0	76.6	3.91
2.5	0.10	0.05	261	45.1	5.78	49.1	74.0	3.63
1.5	0.05	0.10	224	85.9	9.39	130	75.9	9.86
2.0	0.05	0.10	224	85.7	9.39	129	75.6	9.78
2.5	0.05	0.10	223	85.1	9.41	128	74.8	9.55
1.5	0.10	0.10	212	73.5	9.63	95.9	58.4	5.60
2.0	0.10	0.10	211	72.8	9.65	93.8	57.3	5.37
2.5	0.10	0.10	210	71.5	9.67	90.3	55.5	5.01
1–0–3 PZN–0.08PT SC/modified $PbTiO_3$ FC/polyurethane composite								
1.5	0.05	0.05	411	139	5.23	208	252	52.2
2.0	0.05	0.05	410	139	5.23	207	251	51.9
2.5	0.05	0.05	409	137	5.23	205	249	51.1
1.5	0.10	0.05	384	117	5.30	175	220	38.6
2.0	0.10	0.05	382	115	5.30	173	218	37.7
2.5	0.10	0.05	379	113	5.31	169	215	36.3
1.5	0.05	0.10	285	179	8.35	357	162	57.8
2.0	0.05	0.10	284	179	8.36	355	161	57.5
2.5	0.05	0.10	284	177	8.36	353	161	56.7
1.5	0.10	0.10	273	156	8.52	305	145	44.3
2.0	0.10	0.10	272	155	8.53	301	144	43.4
2.5	0.10	0.10	271	152	8.55	295	142	42.0

[a]Electromechanical properties of the 0–3 matrix were evaluated by means of the finite element method [46]

$(Q_{33}^*)^2$, $(Q_{31}^*)^2$ and $(Q_h^*)^2$ are achieved due to the more pronounced piezoelectric anisotropy that is also caused by the elastic properties of the 0–3 matrix. The system of oblate FC inclusions in the softer polymer matrix leads to a large decrease of the transversal piezoelectric response of the composite. As a consequence, irrespective of the piezoelectric activity of the 0–3 matrix, the condition (3.8) holds for the 1–0–3 PZN–0.08PT SC/modified $PbTiO_3$ FC/polyurethane composite (see data in Tables 3.13 and 3.14). The validity of the condition (3.8) and the large values of the squared figures of merit $(Q_{33}^*)^2$, $(Q_{31}^*)^2$ and $(Q_h^*)^2$ are to be taken into account for energy-harvesting applications of the studied 1–0–3 composite.

Table 3.14 Effective parameters calculated for the 1–0–3 PZN–0.08PT SC/modified $PbTiO_3$ FC/polymer composite in a case of a piezo-passive 0–3 matrix[a]

ρ_i	m_i	m	g^*_{33}, mV m/N	$(Q^*_{33})^2$, 10^{-12} Pa^{-1}	$(Q^*_{31})^2$, 10^{-12} Pa^{-1}	d^*_h, pC/N	g^*_h, mV m/N	$(Q^*_h)^2$, 10^{-12} Pa^{-1}
1–0–3 PZN–0.08PT SC/modified $PbTiO_3$ FC/araldite composite								
1.5	0.05	0.05	288	56.9	5.66	72.9	107	7.77
2.0	0.05	0.05	289	57.2	5.66	73.4	107	7.87
2.5	0.05	0.05	289	57.1	5.66	73.3	107	7.85
1.5	0.10	0.05	270	49.2	5.70	58.0	86.3	5.01
2.0	0.10	0.05	271	49.6	5.70	58.9	87.6	5.16
2.5	0.10	0.05	272	49.6	5.70	58.9	87.7	5.16
1.5	0.05	0.10	225	87.0	9.35	133	77.5	10.3
2.0	0.05	0.10	225	87.3	9.34	134	77.9	10.4
2.5	0.05	0.10	225	87.2	9.34	134	77.9	10.4
1.5	0.10	0.10	215	77.1	9.51	107	64.2	6.85
2.0	0.10	0.10	216	77.7	9.85	108	65.1	7.05
2.5	0.10	0.10	216	77.7	9.49	108	65.1	7.04
1–0–3 PZN–0.08PT SC/modified $PbTiO_3$ FC/polyurethane composite								
1.5	0.05	0.05	413	141	5.19	211	255	53.6
2.0	0.05	0.05	414	141	5.18	211	255	53.8
2.5	0.05	0.05	413	141	5.18	211	256	53.8
1.5	0.10	0.05	392	124	5.21	186	231	42.9
2.0	0.10	0.05	393	124	5.20	187	232	43.5
2.5	0.10	0.05	393	124	5.20	187	232	43.3
1.5	0.05	0.10	286	181	8.28	363	163	59.3
2.0	0.05	0.10	286	181	8.27	363	164	59.5
2.5	0.05	0.10	286	181	8.27	363	164	59.4
1.5	0.10	0.10	277	163	8.36	323	151	48.9
2.0	0.10	0.10	277	164	8.34	325	152	49.4
2.5	0.10	0.10	277	164	8.34	324	152	49.3

[a]Electromechanical properties of the 0–3 matrix were evaluated by means of the effective field method [7] at neglecting a piezoelectric activity of FC

3.6 Three Figures of Merit for Transmitter-Receiver or Pulse-Echo Systems

In Sect. 2.6.2 we discussed ECFs and their anisotropy in 1–3–0 composites based on relaxor-ferroelectric SCs. Now we consider a system of squared figures of merit [6, 53] of the 1–3–0 composite as follows:

$$(Q^*_{33})^2 = d^*_{33} g^*_{33}, F^*_{q,33} = (d^*_{33})^2 \text{ and } F^*_{E,33} = (k^*_{33})^4 / \left(Z^*_{a,33}\right)^2 \quad (3.14)$$

for the longitudinal piezoelectric effect (33 oscillation mode), and

$$\left(Q_{3j}^*\right)^2 = d_{3j}^* g_{3j}^*, F_{q,3j}^* = \left(d_{3j}^*\right)^2 \text{ and } F_{E,3j}^* = \left(k_{3j}^*\right)^4 / \left(Z_{a,jj}^*\right)^2 \quad (3.15)$$

for the transversal piezoelectric effect (31 and 32 oscillation modes, $j = 1$ and 2). The squared figures of merit from (3.14) and (3.15) are used to characterise a performance of the piezoelectric transmitter-receiver or pulse-echo system. In (3.14) and (3.15), $Z_{a,ff}^* = \rho^* v_f^*$ is the specific acoustic impedance on either the polar ($f = 3$) or non-polar ($f = 1$ or 2) direction, ρ^* is the density of the composite, and v_f^* is the sound velocity along the co-ordinate axis OX_f.

We analyse the squared figures of merit from (3.14) and (3.15) for the 1–3–0 SC/porous polymer composite described in Sect. 2.6.2. Both the SC rods in the form of a rectangular parallelepiped with a square base and spheroidal air pores in the polymer matrix are distributed regularly. The main component of the 1–3–0 composite is the [001]-poled PMN–0.33PT SC with a high piezoelectric activity (see electromechanical constants in Table 1.1), and the SC rods are surrounded by a piezo-passive porous polyurethane matrix with 3–0 connectivity. Data on the squared figures of merit are shown in Table 3.15 and Fig. 3.14. It is seen that the studied 1–3–0 composite with the system of highly oblate spheroidal air pores in the polymer medium (i.e., at the aspect ratio $\rho_p >> 1$) provides the best set of the parameters from (3.14) and (3.15). The highly oblate pores strongly influence the anisotropy of the elastic properties of the matrix and the piezoelectric response of the composite as a whole: with increasing ρ_p, the transversal piezoelectric effect weakens. As a consequence, conditions

$$\left(Q_{33}^*\right)^2 \gg \left(Q_{31}^*\right)^2 \text{ and } F_{E,33}^* \gg F_{E,31}^* \quad (3.16)$$

are valid in wide ranges of m, m_p and ρ_p.

Maxima of some squared figures of merit at $\rho_p >> 1$ are achieved at relatively small volume fractions of SC m (see Table 3.15 and curve 4 in Fig. 3.14b). However changes in values of the squared figures of merit with increasing m are not considerable (see Fig. 3.14b), and it would be reasonable to manufacture a high-performance composite sample at m = 0.05–0.10. In this volume-fraction range of SC, the influence of the porous polymer matrix on the performance of the composite is still significant. Graphs in Fig. 3.14 contain curves 1 related to the 1–3 composite with the monolithic polymer matrix. It is seen that the squared figures of merit of the 1–3 composite are considerably less than those of the studied 1–3–0 composite, and this effect is caused by the matrix. Thus, the porous polymer matrix improves the set of effective parameters of the 1–3–0 composite that are suitable for piezoelectric transmitter-receiver and/or pulse-echo systems where the 33 oscillation mode is dominant.

Table 3.15 Maximum values of squared figures of merit from (3.14) and (3.15) and related volume fractions of SC m in the 1–3–0 PMN–0.33PT SC/porous polyurethane composite

ρ_p	max $(Q^*_{33})^2$, 10^{-10} Pa^{-1}	max $F^*_{E,33}$, 10^{-14} Rayl^{-2}	max $(Q^*_{31})^2$, 10^{-11} Pa^{-1}	max $F^*_{E,31}$, 10^{-15} Rayl^{-2}
Composite at m_p = 0.1				
1	2.86 (m = 0.088)	12.8 (m = 0.073)	3.74 (m = 0.106)	2.07 (m = 0.077)
10	4.18 (m = 0.056)	20.8 (m = 0.053)	2.50 (m = 0.077)	1.03 (m = 0.054)
100	15.2 (m = 0.014)	91.1 (m = 0.019)	0.446 (m = 0.016)	0.0389 (m = 0.014)
Composite at m_p = 0.2				
1	3.44 (m = 0.070)	17.6 (m = 0.060)	4.20 (m = 0.080)	2.48 (m = 0.060)
10	6.36 (m = 0.035)	37.8 (m = 0.037)	2.21 (m = 0.041)	0.847 (m = 0.032)
100	29.4 (m = 0.008)	204 (m = 0.010)	0.267 (m = 0.007)	0.0143 (m = 0.007)

Notes
1. $F^*_{q,33}$ and $F^*_{q,31}$ are monotonically increasing functions of the volume fraction of SC m irrespective of the parameters m_p and ρ_p of the porous matrix
2. Volume fractions m related to maximum values of the effective parameters in the 2nd–5th columns are given in parentheses
3. The [001]-poled PMN–0.33PT SC is characterised by $(Q^{(1)}_{33})^2 = 1.10 \cdot 10^{-10}$ Pa^{-1}, $F^{(1)}_{q,33} = 7.95 \cdot 10^{-18}$ C^2/N^2, $F^{(1)}_{E,33} = 1.25 \cdot 10^{-14}$ Rayl^{-2}, $(Q^{(1)}_{31})^2 = 0.244 \cdot 10^{-10}$ Pa^{-1}, $F^{(1)}_{q,31} = 1.76 \cdot 10^{-18}$ C^2/N^2, and $F^{(1)}_{E,31} = 0.106 \cdot 10^{-14}$ Rayl^{-2}. The listed parameters were calculated using experimental data from work [16]

3.7 Related Parameters

Now we consider effective parameters that can be regarded as 'relatives' to the figures of merit. If the piezo-composite element radiates acoustic waves, then its specific acoustic power at low frequencies is proportional to $\left(d^*_{ik}/v^{*E}_k\right)^2 E^2$ [54], where d^*_{ik} is the piezoelectric coefficient, $v^{*E}_k = \left(\rho^* s^{*E}_{kk}\right)^{-1/2}$ is the sound velocity measured along OX_k, E is electric field applied to the element and ρ^* is its effective density. It is assumed that frequencies are much less than the resonance frequency of the piezoelectric element. A ratio of the specific acoustic powers related to the longitudinal and transversal piezoelectric responses of the element is given by

$$r^*_{pow} = \left(d^*_{33}/v^{*E}_3\right)^2 / \left(d^*_{31}/v^{*E}_1\right)^2 = \left(\zeta^*_{d3j}\right)^2 s^{*E}_{33}/s^{*E}_{11} \quad (3.17)$$

In (3.17), ζ_{d3j}* is the anisotropy factor concerned with the piezoelectric coefficients d^*_{3j} [see (2.4)], and s^{*E}_{ab} is the elastic compliance at E = const.

Oscillations of the piezoelectric element in the form of the rectangular parallelepiped elongated on the OX_3 direction and conversion of energy in the piezoelectric element can be characterised by coefficients of electromechanical transformation N^*_{33} and N^*_{31} [54, 55] at the longitudinal and transversal oscillation modes, respectively. According to work [54], these coefficients linearly depend on

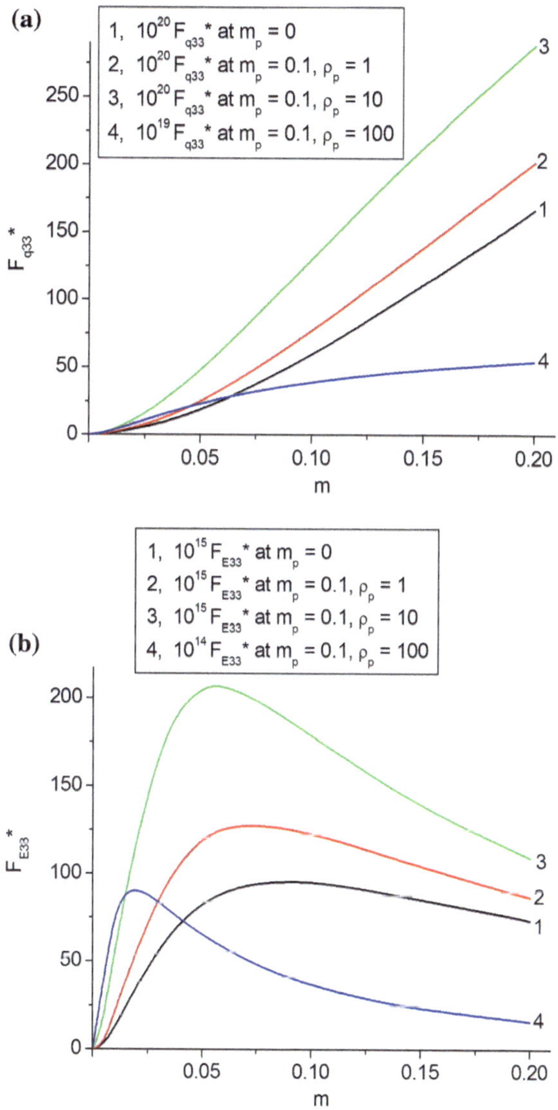

Fig. 3.14 Volume-fraction dependences of squared figures of merit $F_{q,33}^*$ (**a**) and $F_{E,33}^*$ (**b**) of the 1–3 PMN–0.33PT SC/polyurethane composite (at m_p = 0) and 1–3–0 PMN–0.33PT SC/porous polyurethane composite (at m_p = 0.1). The microgeometry of the 1–3–0 SC/porous polymer composite was described in Sect. 2.6.2

d_{3j}^*/s_{jj}^{*E} and the geometric size of the parallelepiped. The ratio of the coefficients of electromechanical transformation of the composite is represented as

$$r_N^* = \left(N_{33}^*/N_{31}^*\right) = \zeta_{d3j}^* s_{33}^{*E}/s_{11}^{*E}. \qquad (3.18)$$

Equations (3.17) and (3.18) suggest that the anisotropy factor $\zeta_{d3j}{}^*$ and the ratio of elastic compliances concerned with longitudinal and transverse responses strongly influence r_{pow} and r_N.

Below we consider examples of the validity of inequalities

$$r^*_{pow} \geq 5 \tag{3.19}$$

and

$$\left| r^*_N \right| \geq 5 \tag{3.20}$$

which are caused by the piezoelectric and elastic anisotropy of the composite. Considerable variations of this anisotropy are observed, for instance, in the 1–3–0 FC/porous polymer composite (Fig. 2.10) with a regular arrangement of FC rods and air inclusions. It is assumed that the remanent polarisation of each FC rod is $\boldsymbol{P}_r^{(1)} \uparrow\uparrow OX_3$, and electrodes on the sample are parallel to the (X_1OX_2) plane (Fig. 2.10). To examine validity of conditions (3.19) and (3.20), we vary the following parameters of the composite: m, m_p and ρ_p.

In Table 3.16 we show volume-fraction (m) ranges that are related to the validity of conditions (3.19) and (3.20) in composites with different FC components and microgeometric characteristics of its porous matrix (m_p and ρ_p). As follows from Table 3.16, the volume-fraction ranges, wherein conditions (3.19) and (3.20) hold, become wider on the transition of a prolate shaped air inclusion in the polymer matrix (see inset in Fig. 2.10) to an oblate shape. This behaviour is a result of the important influence of elastic properties of the porous polymer matrix on composite properties. As follows from our analysis, as the aspect ratio of the air inclusion ρ_p increases at porosity m_p = const the ratios of elastic moduli $c_{33}^{(3)}/c_{13}^{(3)}$ and $c_{11}^{(3)}/c_{13}^{(3)}$ of the porous polymer significantly increase and lead to a more pronounced piezoelectric response of the composite along the OX_3 axis. The important role of the elastic properties of the matrix is additionally confirmed by data on the related 1–3 composites in the end part of Table 3.16. For the monolithic polyurethane matrix, the relation $c_{33}^{(3)}/c_{13}^{(3)} = c_{11}^{(3)}/c_{13}^{(3)} = 1.7$ is valid, and these ratios are approximately equal to the similar ratios related to the FC components considered in Table 3.16.

Following conventional formulae [11, 12] for the piezoelectric medium, we find that the factor from the coefficient of electromechanical transformation N^*_{33} is represented [55] as follows:

$$d^*_{33}/s^{*E}_{33} = \left(2e^*_{31}s^{*E}_{13} + e^*_{33}s^{*E}_{33}\right)/s^{*E}_{33} = 2e^*_{31}\left(s^{*E}_{13}/s^{*E}_{33}\right) + e^*_{33}. \tag{3.21}$$

Because of $\left|s^{*E}_{13}\right| < s^{*E}_{33}$ and the large anisotropy of the piezoelectric coefficients e^*_{3j} in the 1–3-type composite [7, 10], a contribution from $2e^*_{31}\left(s^{*E}_{13}/s^{*E}_{33}\right)$ in (3.21) is small. As a consequence, the equality

$$d^*_{33}/s^{*E}_{33} = e^*_{33} \tag{3.22}$$

holds with an accuracy to 5% in the whole m range [55]. It should be added that the piezoelectric coefficient e^*_{33} linking the piezoelectric polarisation of the composite

Table 3.16 Volume-fraction ranges [m_1; m_2] wherein conditions (3.19) and (3.20) in 1–3-type FC-based composites are valid

ρ_p	m_p	Range [m_1; m_2] of validity of condition (3.19)	Range [m_1; m_2] of validity of condition (3.20)
1–3–0 ZTS-19 FC/porous polyurethane composite			
0.1	0.1	[0.001; 0.011]	[0.032; 0.700]
	0.2	[0.001; 0.003]	[0.018; 0.700]
	0.3	–	[0.010; 0.700]
1	0.1	[0.001; 0.023]	[0.040; 0.700]
	0.2	[0.001; 0.022]	[0.031; 0.700]
	0.3	[0.001; 0.020]	[0.024; 0.700]
10	0.1	[0.001; 0.112]	[0.031; 0.700]
	0.2	[0.001; 0.152]	[0.022; 0.700]
	0.3	[0.001; 0.164]	[0.016; 0.700]
100	0.1	[0.001; 0.321]	[0.007; 0.700]
	0.2	[0.001; 0.325]	[0.004; 0.700]
	0.3	[0.001; 0.312]	[0.003; 0.700]
1–3–0 PZT-5H FC/porous polyurethane composite			
0.1	0.1	[0.001; 0.013]	[0.038; 0.700]
	0.2	[0.001; 0.004]	[0.021; 0.700]
	0.3	–	[0.011; 0.700]
1	0.1	[0.001; 0.029]	[0.049; 0.700]
	0.2	[0.001; 0.037]	[0.037; 0.700]
	0.3	[0.001; 0.024]	[0.028; 0.700]
10	0.1	[0.001; 0.129]	[0.037; 0.700]
	0.2	[0.001; 0.169]	[0.025; 0.700]
	0.3	[0.001; 0.184]	[0.019; 0.700]
100	0.1	[0.001; 0.336]	[0.008; 0.700]
	0.2	[0.001; 0.338]	[0.004; 0.700]
	0.3	[0.001; 0.323]	[0.003; 0.700]
1–3–0 PCR-7M FC/porous polyurethane composite			
0.1	0.1	[0.001; 0.013]	[0.037; 0.700]
	0.2	[0.001; 0.004]	[0.021; 0.700]
	0.3	–	[0.012; 0.700]
1	0.1	[0.001; 0.026]	[0.047; 0.700]
	0.2	[0.001; 0.025]	[0.036; 0.700]
	0.3	[0.001; 0.023]	[0.028; 0.700]
10	0.1	[0.001; 0.120]	[0.036; 0.700]
	0.2	[0.001; 0.157]	[0.025; 0.700]
	0.3	[0.001; 0.168]	[0.019; 0.700]
100	0.1	[0.001; 0.309]	[0.008; 0.700]
	0.2	[0.001; 0.312]	[0.005; 0.700]
	0.3	[0.001; 0.299]	[0.003; 0.700]

(continued)

Table 3.16 (continued)

ρ_p	m_p	Range $[m_1; m_2]$ of validity of condition (3.19)	Range $[m_1; m_2]$ of validity of condition (3.20)
1–3 ZTS-19 FC/polyurethane composite			
–	0	[0.001; 0.023]	[0.054; 0.700]
1–3 PZT-5H FC/polyurethane composite			
–	0	[0.001; 0.029]	[0.065; 0.700]
1–3 PCR-7M FC/polyurethane composite			
–	0	[0.001; 0.027]	[0.064; 0.700]

and the external strain along the poling axis OX_3 [7] shows a weak dependence on elastic properties of the matrix in the 1–3-type composite due to the dominating role of its stiff FC rods oriented along OX_3. This is also illustrated by curve 1 in Fig. 3.15. As for the d_{31}^*/s_{11}^{*E} factor, it undergoes changes with increasing porosity m_p (see curves 2–4 in Fig. 3.15) that can be accounted for by changes in the elastic anisotropy of the porous polymer matrix. Increasing m_p at ρ_p = const leads to a larger elastic anisotropy of the porous matrix and influences the transverse piezoelectric response of the 1–3–0 composite. It should be added that monotonic dependences of d_{3j}^*/s_{jj}^{*E} on the volume fraction m at m_p = const and ρ_p = const are predicted for the 1–3-type composites based on PZT-type FCs, and largest values of d_{3j}^*/s_{jj}^{*E} (j = 1 and 3) are achieved at m = 1, i.e., in the FC.

In contrast to d_{3j}^*/s_{jj}^{*E}, the volume-fraction (m) dependence of the ratio of the coefficients of electromechanical transformation r_N* from (3.18) is non-monotonic for the same 1–3-type composites at m_p = const and ρ_p = const. Data from Table 3.17 show that the elastic anisotropy of the porous matrix influences the ratio of the coefficients of electromechanical transformation r_N* of the 1–3–0 composite.

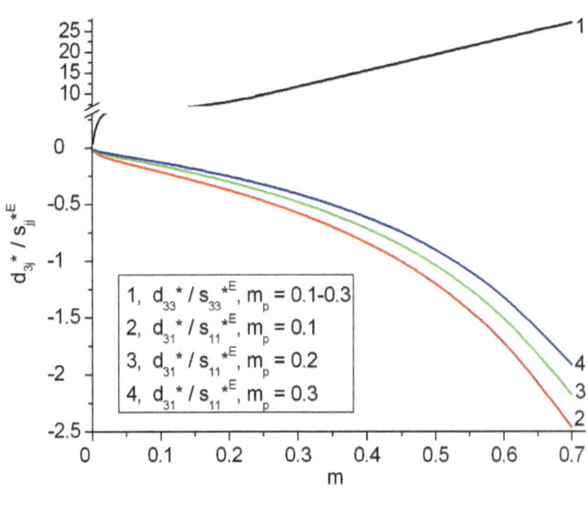

Fig. 3.15 Volume-fraction dependence of d_{3j}^*/s_{jj}^{*E} factors (in C/m^2, j = 1 and 3) related to the 1–3–0 PCR-7M FC/porous polyurethane composite at ρ_p = 100 (reprinted from paper by Topolov et al. [55], with permission from Taylor and Francis)

Table 3.17 Minimum values of $r_N{}^*$ of the 1–3–0 PCR-7M FC/porous polyurethane composite at $m_p = 0.3$ and ratios that characterise the elastic anisotropy of the porous polyurethane matrix

ρ_p	$\min r_N{}^*$	$c_{33}^{(3)}/c_{13}^{(3)}$	$c_{11}^{(3)}/c_{13}^{(3)}$	$\left\lvert c_{33}^{(3)} - c_{11}^{(3)}\right\rvert / c_{13}^{(3)}$
0.1	−20.9 (at m = 0.410)	2.67	0.691	1.98
1	−17.3 (at m = 0.405)	1.70	1.08	0.62
10	−20.4 (at m = 0.323)	1.38	3.09	1.71
100	−31.7 (at m = 0.157)	1.36	29.9	28.5

The change in ρ_p of the porous matrix leads to changes in its elastic moduli $c_{ab}^{(3)}$. With increasing the $\left\lvert c_{33}^{(3)} - c_{11}^{(3)}\right\rvert / c_{13}^{(3)}$ ratio, values of | $\min r_N{}^*$ | increase (see the 2nd and 5th columns in Table 3.17). It is seen that $\min r_N{}^*$ is achieved at volume fractions of FC $0.157 \le m \le 0.410$, i.e., at the dominating volume fraction of porous polymer in the 1–3–0 composite. The revealed correlation (Table 3.17) also testifies to the important role of the elastic anisotropy of the porous polymer matrix in the formation of the piezoelectric performance and parameters (3.17), (3.18), (3.21), and (3.22) of the 1–3–0 composite in the wide volume-fraction (m) range.

The piezoelectric coefficients e_{3j}^* involved in (3.21) and (3.22) are of an independent interest for energy-harvesting applications. These piezoelectric coefficients are proportional to the coupling of electroded cantilevers [56] at one of the oscillations modes, either the '33' or '31' mode. In the 1–3-type composites based on FC or relaxor-ferroelectric SC, the inequality $e_{33}^* \gg \left\lvert e_{31}^*\right\rvert$ is valid within wide volume-fraction (m) ranges, and this condition favours an exploitation of a composite sample at the '33' mode. Since a minor max e_{33}^* appears at $0.9 < m < 1$, and e_{33}^* mainly increases with increasing m [7, 57], the FC (or SC) component becomes more suitable for piezoelectric energy harvesting in comparison to the related composite.

Data from Table 3.18 suggest that the piezoelectric coefficients e_{ij} of the perovskite-type FCs and domain-engineered relaxor-ferroelectric SCs vary in a relatively narrow range, i.e., typical values of $e_{ij} \sim$ (1–10) C/m^2. Positive values of e_{31} and a significant anisotropy of e_{3j} are important features of the $PbTiO_3$-type FCs with the large nisotropy of the piezoelectric coefficients d_{3j} [7, 41]. These characteristics of the $PbTiO_3$-type FCs strongly influence the piezoelectric properties of the related composites with connectivity patterns such as 1–3 [7, 40], 0–3 [7, 49], 2–2 [7], 1–1 [62], and 3–3 [63]. Finally, it should be noted that values of $\lvert e_{3j}\rvert$ of poled domain-engineered relaxor-ferroelectric SCs can be less than $\lvert e_{3j}\rvert$ of some FCs (see Table 3.18) despite the larger piezoelectric coefficients d_{3j} of these SCs. The reason for such a discrepancy is concerned with differences between elastic properties of the FCs and the SCs. It is obvious that these features of the piezoelectric materials are to be taken into account when selecting them for energy-harvesting applications.

Table 3.18 Experimental room-temperature values of piezoelectric coefficients e_{ij} (in C/m^2) of poled FCs and relaxor-ferroelectric SCs [7, 9, 10, 58–61][a]

	e_{31}	e_{33}	e_{15}
FCs[b]			
$BaTiO_3$ (I)	–4.4	18.6	11.6
$BaTiO_3$ (II)	–4.35	17.5	11.4
$(Ba_{0.917}Ca_{0.083})TiO_3$	–3.14	13.5	10.9
TBK-3	–3.2	12.5	9.1
TBKS	–0.68	7.71	4.56
NBS-1	–3.3	14.8	7.7
ZTS-19	–4.9	14.9	10.6
ZTS-24	–3.54	12.5	10.3
ZTSNV-1	–8.14	15.9	14.3
PZT-4	–5.2	15.1	12.7
PZT-5	–5.4	15.8	12.3
PZT-5H	–6.5	23.3	17.0
PZT-7A	–2.3	9.3	9.2
Navy Type VI	–6.5	23.3	17.0
PZ 27	–3.71	15.77	11.74
PZ 34	3.81	6.87	2.31
PCR-1, hp[c]	–3.7	11.1	10.8
PCR-7, hp	–9.0	28.3	17.9
PCR-7M, hp	–9.5	31.1	20.0
PCR-8, hp	–7.5	13.6	11.6
PCR-8, ct	–6.9	15.2	11.6
PCR-13, ct	–6.0	10.4	7.1
PCR-21, hp	–7.3	16.2	12.0
PCR-63, hp	–4.6	11.8	6.9
PCR-73, hp	–9.1	29.2	22.4
$Pb(Zr_{0.54}Ti_{0.46})O_3$	–1.86	9.0	9.8
$Pb(Zr_{0.52}Ti_{0.48})O_3$	–3.08	11.0	10.2
$(Pb_{0.94}Sr_{0.06})(Ti_{0.47}Zr_{0.53})O_3$	–5.2	15.1	12.7
Modified $PbTiO_3$ (I)	0.4	6.7	3.0
$(Pb_{0.9625}La_{0.025})(Ti_{0.99}Mn_{0.01})O_3$	1.32	6.80	3.35
$(Pb_{0.85}Nd_{0.10})(Ti_{0.99}Mn_{0.01})O_3$	1.80	8.83	5.41
$(Pb_{0.855}Nd_{0.11})(Ti_{0.94}Mn_{0.02}In_{0.04})O_3$	2.00	8.81	4.84
PMN– 0.35PT	–5.0	14.3	28.0
Domain-engineered [001]-poled SCs[d]			
PMN–0.30PT	–2.4	27.1	13.6
PMN–0.33PT	–3.9	20.3	10.1
PZN–0.08PT	–5.1	15.4	10.1

[a]See also references in the 1st column of Table 1.2
[b]∞mm symmetry
[c]FC samples of the PCR type have been manufactured using either the conventional technology (ct) or hot pressing (hp). PCR is the abbreviation for the group 'piezoelectric ceramics from Rostov-on-Don' (Russia) [7, 9]
[d]Macroscopic 4*mm* symmetry

It should be added that the shear oscillation mode can be of an independent interest for piezoelectric energy-harvesting applications. In work [64] a vibration energy harvesting device using the shear mode of the PMN–0.29PT SC has been designed and fabricated. In this device the high performance is achieved due to large values of the squared figure merit $d_{15}g_{15}$ concerned with the significant shear piezoelectric effect. According to experimental data [64], the studied domain-engineered SC is characterised by values of $d_{15}g_{15} = (261–267) \cdot 10^{-12}$ Pa^{-1}. These values are, for instance, by about 9 times larger than the squared figure of merit $(Q_{33})^2 = d_{33}g_{33} = 28.8 \cdot 10^{-12}\ Pa^{-1}$ related to the longitudinal piezoelectric effect in the domain-engineered PMN–0.28PT SC, a sample with the closest chemical composition (see data in Table 1.3). The $d_{15}g_{15}$ values are also by a few times larger than $(Q_{33})^2$ of the remaining PMN–xPT and PZN–yPT SCs listed in Table 1.3. It is believed that the shear oscillation mode and related effective parameters in piezo-active composites and their potential components will be a subject of studies in the future.

3.8 Conclusion

Chapter 3 has been devoted to the system of squared figures of merit that are concerned with the electromechanical properties of the piezoelectric medium and used to characterise its performance in the context of transducer, hydroacoustic and energy-harvesting applications. We have considered a number of examples of the squared figures of merit in conventional poled FCs and piezo-active composites based on either FCs or relaxor-ferroelectric SCs. The main results discussed in Chap. 3 are formulated as follows.

(i) The squared figures of merit have been analysed for poled FCs with the perovskite-type structure, piezo-active composites based on these FCs and piezo-active composites based on relaxor-ferroelectric SCs with a high piezoelectric activity. Among the connectivity patterns of interest for piezoelectric energy-harvesting and related applications, we mention 2–2, 1–3, 0–3, 1–0–3, and 1–3–0 systems.

(ii) The 2–2 PZT-type FC/polymer composite with a parallel connection of layers is of interest due to squared figures of merit $\left[Q^*_{3j}(m)\right]^2 \gg \left(Q^{(1)}_{3j}\right)^2$ (see Table 3.2), where the squared figures of merit $\left(Q^{(1)}_{3j}\right)^2$ (j = 1, 2 and 3) are related to FC.

(iii) The class of 2–2 parallel-connected composites based on PMN–0.28PT SCs with high piezoelectric activity and three variants of the poling direction of these SCs have been studied in detail. Examples of the polarisation orientation effects are related to the 2–2 composites with the SC components poled along one of the directions - [111], [001] or [011] in the perovskite unit cell. The squared hydrostatic figure of merit $(Q_h^*)^2$ from (3.6) strongly depends on the rotation mode of the main crystallographic axes in the SC layer. For the 2–2 composite based on the [011]-poled PMN–0.28PT SC, the largest $(Q_h^*)^2$ value is achieved (Table 3.6) at the simplest orientation of the main crystallographic axes.

(iv) An additional opportunity to increase $(Q_h^*)^2$ is concerned with replacing the monolithic polymer component with an auxetic polymer. For instance, in the 2–2-type PIN-0.24-0.49 SC/auxetic polyethylene composite, the $(Q_h^*)^2$ value varies from about $150 \cdot 10^{-12}$ Pa^{-1} to $70 \cdot 10^{-12}$ Pa^{-1} at volume fractions of SC $m = 0.05$–0.10 (see Fig. 3.4).

(v) Condition (3.8) for the large anisotropy of the squared figures of merit $\left(Q_{3f}^*\right)^2$ in 2–2-type PMN–xPT/auxetic polyethylene composites is a result of the polarisation orientation effect, elastic anisotropy, and the relatively high piezoelectric activity of the SC component.

(vi) The 1–3-type PCR-7M FC/auxetic polyethylene composite is of interest due to large values of $(Q_{33}^*)^2$ (Table 3.8). For this composite system, a relatively large volume-fraction range is found wherein condition (3.8) $(Q_{3j}^*)^2$ holds. This is due to the active role of the auxetic component in forming the piezoelectric response of the composite.

(vii) The lead-free 1–3-type SC/auxetic polyethylene composites based on either KNN-TL or KNN-T SCs are advanced materials with high piezoelectric sensitivity [$g_{33}^* \sim (10^2$–$10^3)$ mV m/N], large squared figures of merit [$(Q_{33}^*)^2 \sim 10^{-11}$ Pa^{-1} and $(Q_h^*)^2 \sim (10^{-11}$ –$10^{-10})$ Pa^{-1}] (Fig. 3.9), and change in the sign of the piezoelectric coefficient d_{31}^*. This is due to the relatively high piezoelectric activity of the [001]-poled SC and elastic properties of the auxetic polymer component.

(viii) Examples of the 0–3-type composites based on the [001]-poled PMN – 0.33PT SC suggest that squared figures of merit $(Q_{3j}^*)^2$ and $(Q_h^*)^2$ strongly depend on the volume fraction of SC m and on the aspect ratio of SC inclusions ρ therein (see Table 3.9 and Fig. 3.10).

(ix) The new orientation effect in 1–0–3 composites with two contrasting ferroelectric components [i.e., highly piezo-active PMN–0.33PT SC (rods) and highly anisotropic $(Pb_{1-x}Ca_x)TiO_3$ FC (inclusions in a matrix, $x = 0.20$–0.25)] has been studied to show the high performance of these materials. We show the influence of the mutual orientation of the poling direction of SC and FC components on the properties of a 1–0–3 PMN–0.33PT SC/$(Pb_{1-x}Ca_x)TiO_3$ FC/araldite composite with $x = 0.20$–0.25. The elastic and piezoelectric anisotropy of the 0–3 FC/polymer matrix with prolate inclusions leads to values of $(Q_h^*)^2 \sim 10^{-11}$ Pa^{-1} in the aforementioned composite within specific volume-fraction and rotation-angle ranges due to a new orientation effect in the presence of a highly anisotropic 0–3 matrix.

The orientation effect leads also to a considerable anisotropy of squared figures of merit, and condition (3.8) holds within specific volume-fraction and rotation-angle ranges.

(x) The influence of the aspect ratio of FC inclusions ρ_i on the piezoelectric performance and hydrostatic parameters of the 1–0–3 composite based on [001]-poled PMN–0.33PT SC has been studied. A significant feature of this composite is that elastic properties of its 0–3 matrix strongly depend on ρ_i and influence the effective electromechanical properties of the 1–0–3 composite in wide aspect-ratio and volume-fraction ranges. Large values of the squared figures of merit $(Q_h^*)^2 \sim 10^{-11}\,\text{Pa}^{-1}$ and $(Q_{33}^*)^2 \sim 10^{-10}\,\text{Pa}^{-1}$ (Table 3.11 and Fig. 3.13b) and the validity of condition (3.13) make this composite attractive in hydrophone and piezoelectric energy-harvesting applications.

(xi) The important role of the elastic properties of the 0–3 FC/polymer matrix in forming the piezoelectric properties and figures of merit of the 1–0–3 PZN–0.08PT SC/modified $PbTiO_3$ FC/polyurethane composite has been discussed. Despite small negative values of the hydrostatic piezoelectric coefficients $d_h^{(1)}$ and $g_h^{(1)}$ of the SC component, this composite system is of interest due to large values of $(Q_{33}^*)^2$, $(Q_{31}^*)^2$ and $(Q_h^*)^2$ (Tables 3.13 and 3.14) which are achieved due to the pronounced piezoelectric anisotropy caused by the elastic properties of the 0–3 matrix with oblate FC inclusions. Moreover, condition (3.8) is valid for the 1–0–3 PZN–0.08PT SC/modified $PbTiO_3$ FC/polyurethane composite, i.e., in the presence of a polymer component with larger elastic compliances.

(xii) The system of figures of merit from (3.14) and (3.15) has been analysed for 1–3-type composites based on relaxor-ferroelectric SCs. The role of the porous polymer matrix in forming the piezoelectric response and anisotropy of squared figures of merit (see conditions 3.16) has been discussed. Large values of $(Q_{33}^*)^2$ and $F_{E,33}^*$ in comparison to those of the SC component (see Table 3.14 and Fig. 3.15) suggest that the 1–3–0 PMN–0.33PT SC/porous polyurethane composite at relatively small porosity (m_p = 0.1–0.2) in the polymer matrix can be effectively used in piezoelectric energy-harvesting applications.

(xiii) Effects of the porosity and microgeometry of the porous polymer matrix on the volume-fraction behaviour of the effective parameters and anisotropy factors in the 1–3-type composite have been analysed in the context of the elastic anisotropy of the matrix. Among the parameters to be of interest for energy-harvesting applications, we mention coefficients of electromechanical transformation and specific acoustic power [see (3.17) and (3.18)]. Their anisotropy considerably depends on the piezoelectric and elastic properties of the composite, and these properties can be varied due to changes in the microgeometry and porosity of the matrix. The correlation between the elastic anisotropy of the porous polymer matrix and minima of the ratio of specific acoustic powers N_{33}^*/N_{31}^* (Table 3.17) has been stated.

In general, knowledge of the figures of merit and related parameters as well as their dependences on microgeometry and content of the piezo-active composite is of benefit for design of novel high-effective piezoelectric energy-harvesting materials with preferable directions for the conversion and propagation of energy along specific directions.

References

1. Grekov AA, Kramarov SO, Kuprienko AA (1989) Effective properties of a transversely isotropic piezoelectric composite with cylindrical inclusions. Mech Compos Mater 25:54–61
2. Kim H, Tadesse Y, Priya S (2009) Piezoelectric energy harvesting. In: Priya S, Inman DJ (eds) Energy harvesting technologies. Springer, New York, pp 3–39
3. Uchino K, Ishii T (2010) Energy flow analysis in piezoelectric energy harvesting systems. Ferroelectrics 400:305–320
4. Priya S (2010) Criterion for material selection in design of bulk piezoelectric energy harvesters. IEEE Trans Ultrason Ferroelectr Freq Control 57:2610–2612
5. Bowen CR, Taylor J, Le Boulbar E, Zabeka D, Topolov VYu (2015) A modified figure of merit for pyroelectric energy harvesting. Mater Lett 138:243–246
6. Sessler GM, Hillenbrand J (2013) Figure of merit of piezoelectret transducers for pulse-echo or transmit-receive systems for airborne ultrasound. Appl Phys Lett 103:122904
7. Topolov VYu, Bowen CR (2009) Electromechanical properties in composites based on ferroelectrics. Springer, London
8. Safari A, Akdogan EK (eds) (2008) Piezoelectric and acoustic materials for transducer applications. Springer, New York
9. Dantsiger AYa, Razumovskaya ON, Reznitchenko LA, Grineva LD, Devlikanova RU, Dudkina SI, Gavrilyatchenko SV, Dergunova NV, Klevtsov AN (1994) Highly effective piezoceramic materials (Handbook). Kniga, Rostov-on-Don (in Russian)
10. Topolov VYu, Bisegna P, Bowen CR (2014) Piezo-active composites. Orientation effects and anisotropy factors. Springer, Heidelberg
11. Ikeda T (1990) Fundamentals of piezoelectricity. Oxford University Press, Oxford
12. Zheludev IS (1971) Physics of crystalline dielectrics, vol 2. Electrical properties. Plenum, New York
13. Gururaja TR, Safari A, Newnham RE, Cross LE (1988) Piezoelectric ceramic-polymer composites for transducer applications. In: Levinson M (ed) Electronic ceramics: properties, devices, and applications. Marcel Dekker, New York Basel, pp 92–128
14. Akdogan EK, Allahverdi M, Safari A (2005) Piezoelectric composites for sensor and actuator applications. IEEE Trans Ultrason Ferroelectr Freq Control 52:746–775
15. Safari A, Akdogan EK (2006) Rapid prototyping of novel piezoelectric composites. Ferroelectrics 331:153–179
16. Zhang R, Jiang B, Cao W (2001) Elastic, piezoelectric, and dielectric properties of multidomain $0.67Pb(Mg_{1/3}Nb_{2/3})O_3$–$0.33PbTiO_3$ single crystals. J Appl Phys 90:3471–3475
17. Fu H, Cohen RE (2000) Polarization rotation mechanism for ultrahigh electromechanical response in single-crystal piezoelectrics. Nature 403:281–283
18. Peng J, Luo H, He T, Xu H, Lin D (2005) Elastic, dielectric, and piezoelectric characterization of $0.70Pb(Mg_{1/3}Nb_{2/3})O_3$–$0.30PbTiO_3$ single crystal. Mater Lett 59:640–643
19. Wang F, He C, Tang Y (2007) Single crystal $0.7Pb(Mg_{1/3}Nb_{2/3})O_3$–$0.3PbTiO_3$/epoxy 1–3 piezoelectric composites prepared by the lamination technique. Mater Chem Phys 105:273–277
20. Ren K, Liu Y, Geng X, Hofmann HF, Zhang QM (2006) Single crystal PMN–PT/epoxy 1–3 composite for energy-harvesting application. IEEE Trans Ultrason Ferroelectr Freq Control 53:631–638

21. Topolov VYu, Krivoruchko AV (2009) Polarization orientation effect and combination of electromechanical properties in advanced $0.67Pb(Mg_{1/3}Nb_{2/3})O_3$–$0.33PbTiO_3$ single crystal/polymer composites with 2–2 connectivity. Smart Mater Struct 18:065011
22. Liu G, Jiang W, Zhu J, Cao W (2011) Electromechanical properties and anisotropy of single- and multi-domain $0.72Pb(Mg_{1/3}Nb_{2/3})O_3$–$0.28PbTiO_3$ single crystals. Appl Phys Lett 99:162901
23. Kar-Gupta R, Venkatesh TA (2008) Electromechanical response of piezoelectric composites: Effects of geometric connectivity and grain size. Acta Mater 56:3810–3823
24. Zhang R, Jiang B, Cao W (2003) Single-domain properties of $0.67Pb(Mg_{1/3}Nb_{2/3})O_3$–$0.33PbTiO_3$ single crystals under electric field bias. Appl Phys Lett 82:787–789
25. Cao HV, Schmidt H, Zhang R, Cao W, Luo H (2004) Elastic, piezoelectric, and dielectric properties of $0.58Pb(Mg_{1/3}Nb_{2/3})O_3$–$0.42PbTiO_3$ single crystal. J Appl Phys 96:549–554
26. Sessler G (1981) Piezoelectricity in polyvinylidenefluoride. J Acoust Soc Am 70:1596–1608
27. Topolov VYu, Bowen CR, Krivoruchko AV (2013) Role of domain orientations in forming the hydrostatic performance of novel 2–2 single crystal/polymer composites. Ferroelectrics 444:84–89
28. Grekov AA, Kramarov SO, Kuprienko AA (1987) Anomalous behavior of the two-phase lamellar piezoelectric texture. Ferroelectrics 76:43–48
29. Topolov VYu, Krivoruchko AV, Bowen CR (2012) Anisotropy of electromechanical properties and hydrostatic response of advanced 2–2-type composites. Phys Status Solidi A 209:1334–1342
30. Sun E, Cao W, Jiang W, Han P (2011) Complete set of material properties of single domain $0.24Pb(In_{1/2}Nb_{1/2})O_3$-$0.49Pb(Mg_{1/3}Nb_{2/3})O_3$-$0.27PbTiO_3$ single crystal and the orientation effects. Appl Phys Lett 99:032901
31. Groznov IN (1983) Dielectric permittivity. In: Physics encyclopaedia. Sovetskaya Entsiklopediya, Moscow, pp 178–179 (in Russian)
32. Bowen CR, Topolov VYu (2014) Polarisation orientation effects and hydrostatic parameters in novel 2–2 composites based on PMN-xPT single crystals. Ferroelectrics 466:21–28
33. Topolov VYu, Filippov SE, Bisegna P (2012) Anisotropy factors and hydrostatic parameters of 1–3-type piezo-active composites with auxetic polymer matrices. Ferroelectrics 432:92–102
34. Topolov VYu, Panich AE (2009) Problem of piezoelectric sensitivity of 1–3-type composites based on ferroelectric ceramics. Ferroelectrics 392:107–119
35. Topolov VYu, Turik AV (2001) Porous piezoelectric composites with extremely high reception parameters. Tech Phys 46:1093–1100
36. Dunn ML, Taya M (1993) Micromechanics predictions of the effective electroelastic moduli of piezoelectric composites. Int J Solids Struct 30:161–175
37. Dunn ML, Taya M (1993) An analysis of piezoelectric composite materials containing ellipsoidal inhomogeneities. Proc R Soc (Lond), Pt A 443:265–287
38. Dunn ML, Wienecke HA (1997) Inclusions and inhomogeneities in transversely isotropic piezoelectric solids. Int J Solids Struct 34:3571–3582
39. Huang JH, Yu S (1994) Electroelastic Eshelby tensors for an ellipsoidal piezoelectric inclusion. Compos Eng 4:1169–1182
40. Topolov VYu, Bisegna P, Krivoruchko AV (2008) Features of electromechanical properties of 1–3 composites based on $PbTiO_3$-type ceramics. J Phys D Appl Phys 41:035406
41. Glushanin SV, Topolov VYu, Krivoruchko AV (2006) Features of piezoelectric properties of 0–3 $PbTiO_3$-type ceramic/polymer composites. Mater Chem Phys 97:357–364
42. Bezus SV, Topolov VYu, Bowen CR (2006) High-performance 1–3-type composites based on $(1-x)$ $Pb(A_{1/3}Nb_{2/3})O_3$–$x PbTiO_3$ single crystals (A = Mg, Zn). J Phys D Appl Phys 39:1919–1925
43. Topolov VYu, Krivoruchko AV, Bisegna P, Bowen CR (2008) Orientation effects in 1–3 composites based on $0.93Pb(Zn_{1/3}Nb_{2/3})O_3$–$0.07PbTiO_3$ single crystals. Ferroelectrics 376:140–152
44. Topolov VYu, Bowen CR, Bisegna P, Krivoruchko AV (2015) New orientation effect in piezo-active 1–3-type composites. Mater Chem Phys 151:187–195

45. Topolov VYu, Bowen CR, Bisegna P (2015) New aspect-ratio effect in three-component composites for piezoelectric sensor, hydrophone and energy-harvesting applications. Sens Actuators A—Phys 229:94–103
46. Topolov VYu, Bowen CR, Bisegna P, Panich AE (2015) Effect of the matrix subsystem on hydrostatic parameters of a novel 1–3-type piezo-composite. Funct Mater Lett 8:1550049
47. Topolov VYu, Bowen CR (2015) High-performance 1–3-type lead-free piezo-composites with auxetic polyethylene matrices. Mater Lett 142:265–268
48. Bisegna P (2015) Private communication
49. Topolov VYu, Bisegna P, Bowen CR (2011) Analysis of the piezoelectric performance of modern 0–3-type composites based on relaxor-ferroelectric single crystals. Ferroelectrics 413:176–191
50. Xu Y (1991) Ferroelectric Materials and their Applications. North-Holland, Amsterdam
51. Topolov VYu, Bowen CR, Filippov SE (2012) High performance of novel 1–3-type composites based on ferroelectric PZT-type ceramics. Ferroelectrics 430:92–97
52. Zhang S, Li F (2012) High performance ferroelectric relaxor-$PbTiO_3$ single crystals: Status and perspective. J Appl Phys 111:031301
53. Topolov VYu, Bowen CR, Isaeva AN (2016) Figures of merit and related parameters of modern piezo-active 1–3-type composites for energy-harvesting applications. In: Parinov IA, Chang S-H, Topolov VYu (eds) Proceedings of the 2015 International conference on Physics, mechanics of new materials and their applications. Nova, New York (in press)
54. Sverdlin GM (1990) Applied Hydroacoustics. Soodostroyeniye, Leningrad (in Russian)
55. Topolov VYu, Filippov SE, Panich AE, Panich EA (2014) Highly anisotropic 1–3–0 composites based on ferroelectric ceramics: microgeometry—volume-fraction relations. Ferroelectrics 460:123–137
56. Kim M, Kim S-H, Hong S (2013) Materials and devices for MEMS piezoelectric energy harvesting. In: Elvin N, Erturk A (eds) Advances in energy harvesting methods. Springer, New York, pp 417–435
57. Topolov VYu, Glushanin SV (2002) Evolution of connectivity patterns and links between interfaces and piezoelectric properties of two-component composites. J Phys D Appl Phys 35:2008–2014
58. Topolov VYu, Panich AE (2008) Electromechanical properties of poled ferroelectric ceramics based on peroskite-family oxides. Issledovano v Rossii, Reg. No. 002, pp 8–26 (in Russian). http://zhurnal.ape.relarn.ru/articles/2008/002.pdf
59. Khoroshun LP, Maslov BP, Leshchenko PV (1989) Prediction of effective properties of piezo-active composite materials. Naukova Dumka, Kiev (in Russian)
60. Luchaninov AG (2002) Piezoelectric effect in non-polar heterogeneous ferroelectric materials. Volgograd State Academy of Architecture and Construction, Volgograd (in Russian)
61. Nagatsuma K, Ito Y, Jyomura S, Takeuchi H, Ashida S (1985) Elastic properties of modified $PbTiO_3$ ceramics with zero temperature coefficients. In: Taylor GW (ed) Ferroelectricity and Related Phenomena, vol 4., PiezoelectricityGordon and Breach Science Publishers, New York, pp 167–176
62. Glushanin SV, Topolov VYu (2001) Features of electromechanical properties of piezoelectric composites with elements of connectivity 1–1. J Phys D Appl Phys 34:2518–2529
63. Bowen CR, Topolov VYu (2003) Piezoelectric sensitivity of $PbTiO_3$-based ceramic/polymer composites with 0–3 and 3–3 connectivity. Acta Mater 51:4965–4976
64. Ren B, Or SW, Wang F, Zhao X, Luo H, Li X, Zhang Q, Di W, Zhang Y (2010) Piezoelectric energy harvesting based on shear mode $0.71Pb(Mg_{1/3}Nb_{2/3})O_3$-$0.29PbTiO_3$ single crystals. IEEE Trans Ultrason Ferroelectr Freq Control 57:1419–1425

Chapter 4
Piezoelectric Mechanical Energy Harvesters and Related Materials

4.1 Examples of Energy-Harvesting Systems

Energy harvesting (or 'energy scavenging') is a subject that continues to receive both industrial and academic interest since it provides a route to create autonomous and self-powered low-power electronic devices, example applications include wireless sensor networks. An excellent commercial example is the recent commercial system developed which converts the vibration of rolling stock into electrical power for the wireless communication of sensors that predict the failure of rail wheel bearings [1]. The ability to deliver sustainable power to a wireless system network by energy harvesting is attractive since it removes the cost of DC batteries and it also reduces the time and cost that is necessary to replace and maintain the batteries. The labour costs that are necessary to install complex wired systems are also avoided. This is particularly relevant to the installation of sensors that are deployed in areas that are either inhospitable or difficult to reach; example includes safety-monitoring devices [1], structure-embedded micro-sensors and medical implants. There are also environmental benefits associated with limiting the use and subsequent disposal of batteries.

Energy-harvesting devices therefore provide a 'battery-less' solution by scavenging energy from ambient sources of energy such as mechanical vibrations, heat, light, water etc., and converting it into a useable form, often electrical power [2]. While energy-harvesting technologies have been continuously improving in the last years, there are also parallel improvements in microprocessor technology leading to an increase in power efficiency and reduced power consumption. Local electrical energy storage solutions are also improving: we mention, for example, the development of super-capacitors [3] and even 'structural power' [4]. It is the successful convergence of harvesting, storage and electronics that will ultimately lead to successful energy-harvesting products and systems.

C.R. Bowen et al., *Modern Piezoelectric Energy-Harvesting Materials*,
Springer Series in Materials Science 238, DOI 10.1007/978-3-319-29143-7_4

As a result of its topical nature, there are already a number of excellent reviews in the area of energy harvesting, which often concentrate on modern nano-scale materials and devices ('nano-generators') [5–13] or surveys of the various potential devices and systems [14–19]. In this chapter we discuss examples of energy harvesters and piezoelectric materials that can be used as active elements within such systems.

4.1.1 Energy-Harvesting Materials and Systems: General Characteristics

Piezoelectric materials are attractive for a number of applications concerned with energy conversion and harvesting. This includes the potential to convert mechanical vibrations into electrical energy via the direct piezoelectric effect, the conversion of thermal fluctuations into electrical energy via the pyroelectric effect and the prospect of using the internal electric fields present in ferroelectrics or strained piezoelectric materials to influence electron-hole recombination in solar-cell devices or chemical reactions, such as water-splitting. The aforementioned materials belong to so-called active dielectrics, and their physical properties are often exploited in external electric, mechanical or thermal fields. Figure 4.1 shows the relationships between the active dielectrics (see the left part of the diagram) and links between the active and passive dielectrics. Among the active dielectrics, ferroelectrics are of particular interest because of the combination of pyroelectric and piezoelectric properties and because of an influence of additional factors (e.g. relaxor, semiconductor or ferroelastic characteristics of some ferroelectrics) on these properties. It is probable that the left part of the diagram shown in Fig. 4.1 may be modified in the future due to a more detailed classification of ferroelectrics and related materials. As follows from Fig. 4.1, all pyroelectrics are piezoelectric and all ferroelectrics are piezoelectric and pyroelectric. Since the aforementioned properties are, in many cases, present in the same material, it provides the intriguing prospect of a material that can harvest energy from multiple sources including, for instance, vibration, sound waves, thermal fluctuations, and light.

Mechanical harvesting, which is the focus of the present monograph, converts energy from movement or vibration into electrical energy. There are a wide variety of sources of mechanical energy including vibrations from industrial machinery and transport [1], fluid flow such as air movements [20, 21], direct human action from walking [22], or in-body motion such as chest and heart movement to power pacemakers [23] and orthopaedic implants [22]. Many of these sources are also used for large-scale power generation, e.g. wind, but energy-harvesting technologies are predominately focussed on small-scale power generation (as a rule, from 10^{-6} W to 10^{-3} W) at the point of use, typically to power small electronic devices where mains or battery power does not provide a viable or convenient solution. In general there are

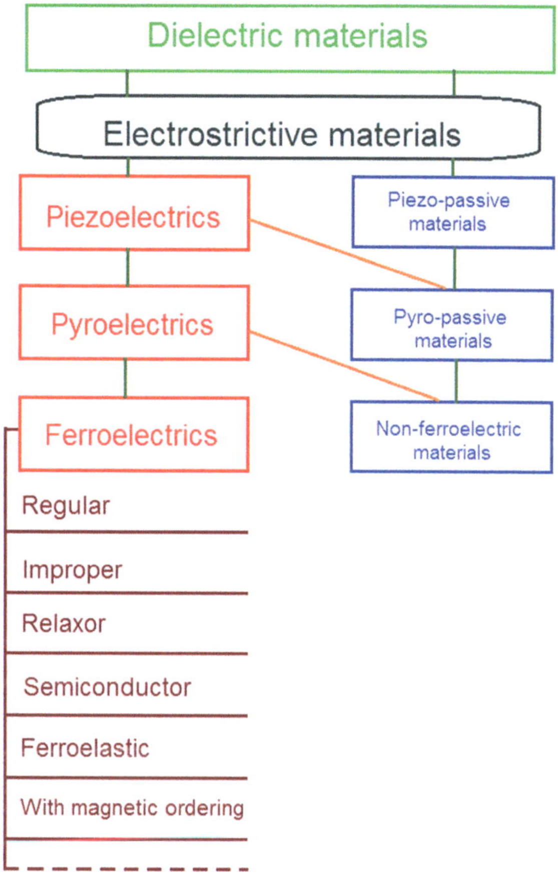

Fig. 4.1 Relationships between dielectric materials

two main ways of extracting energy from a mechanical source. These main ways are described in this chapter as '*inertial*' and '*kinematic*' [24].

Inertial energy harvesting relies on the resistance to acceleration of a mass. This generates a force in a mass-spring system when the source moves. These systems are commonly used for vibration harvesting systems that are connected to the source at a single point [24]. When the source moves, a vibration is set up in the mass-spring system and electrical energy can be extracted. The amplitude of the vibration is not simply related to the source amplitude since the vibration amplitude of a system, if operating at resonance, can be significantly larger than the amplitude of the source movement.

Kinematic energy harvesting couples the energy harvester directly to the relative movement of different parts of the source. Examples of kinematic energy harvesters include harvesting energy from the bending of a tyre wall to power a tyre pressure sensor [24], or the motion of human limbs to power low power electronics [25]. These energy-harvesting mechanisms do not rely on inertia or resonance as in

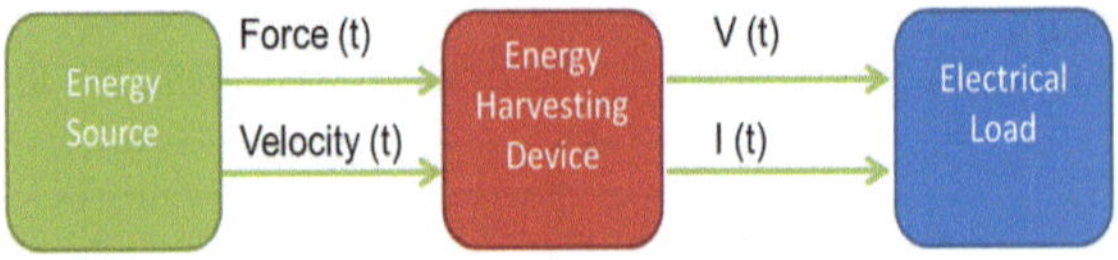

Fig. 4.2 Schematic diagram of an energy-harvesting system. $V(t)$ and $I(t)$ are dependent on voltage V and current I at time t (reproduced from paper by Bowen et al. [24], with permission from The Royal Society of Chemistry)

inertial energy harvesting. Since the strain in the harvester is coupled to flexure or extension of the source, they are connected at more than one point.

An electro-mechanical energy harvester extracts energy from the motion of a source using one of the mechanisms described above and converts it to electrical energy that is delivered to an electrical load. An example of the schematic diagram is shown in Fig. 4.2.

To fully understand the operation of an energy harvester, it is useful to know:

(i) the characteristics of the energy source,
(ii) how energy is transferred from the source to the energy harvester,
(iii) the nature of the electromechanical conversion in the energy-harvesting transducer (e.g. piezoelectric), and
(iv) how energy is transferred from the energy harvester to the electrical load.

Losses can be incurred in the system shown in Fig. 4.2, not just within the energy-harvesting transducer [26], but at all stages in this process. For vibration harvesting, the simplest vibration source is a single frequency sinusoid which is characterised by its frequency and amplitude. The amplitude is commonly defined by the acceleration in energy harvesting applications, but it could also be defined by the velocity or the displacement since they are simply related through the frequency of vibration [24] in accordance with formulae

$$a(t) = -a_0 \sin(2\pi ft), v(t) = v_0 \cos(2\pi ft) \text{ and } l(t) = l_0 \sin(2\pi ft),$$

where a, v and l are acceleration, velocity and distance, respectively, a_0, $v_0 = a_0/(2\pi f)$ and $l_0 = a_0/(2\pi f)^2$ are their amplitudes, f is the frequency, and t is time. In real applications, most vibration sources are rarely simple sinusoids, but there are a number of sources, e.g. machinery operating at AC mains frequency, which have a strong frequency component at a frequency accessible to energy-harvesting devices. Typical vibration sources with an identifiable frequency peak have been characterised [27] by their amplitude at their fundamental mode; for example, producing an acceleration of 3 $m \cdot s^{-2}$ at 13 Hz for a car instrument panel or an acceleration of 12 $m \cdot s^{-2}$ at 200 Hz for a car engine compartment. However, many vibration sources cannot be characterised in this way. For this reason, the energy-harvester performance in complex and broadband vibrational environments must largely be evaluated empirically.

There are a number of transduction technologies for using motion to generate electrical power. Electromagnetic generators are a well-established means of converting mechanical to electrical energy and have been deployed for vibration energy harvesting [1]. These technologies use established manufacturing and engineering methods and are effective both in terms of cost and performance at sizes from a few cubic centimetres upwards. However, performance and ease of manufacture falls rapidly at smaller length scales, so this technology is generally unsuitable for small-scale energy-harvesting applications (e.g. thin sections with areas approximately 1 cm^2 or less). Piezoelectric energy-harvesting materials, the focus of the present monograph, provide a route for solid-state conversion between electrical and mechanical energy, and the materials can be manufactured at small scales and integrated into micro-scale devices and electronic circuits. Power density for piezoelectric transduction exceeds that for electromagnetic generators below around 0.5 $W{\cdot}cm^{-3}$ [28].

To produce electrical energy, the piezoelectric material must be able to generate both charge and voltage. Many piezoelectric materials of technological importance for energy-harvesting applications possess a well-defined polar axis, and the energy-harvesting performance depends on the direction of the applied strain (or stress) relative to this polar axis. In a poled FC or ferroelectric polymer, the polar axis is the poling direction, whilst for non-ferroelectric crystalline materials such as zinc oxide (ZnO), gallium nitride (GaN) or aluminium nitride (AlN), this is defined by the orientation of the crystallographic axes. In these cases, the polar axis is referred to as the '*3*' direction. By symmetry all directions in the plane at right angles to the polar axis are equivalent and are referred to as the '1' direction; this is typical for most ceramic piezoelectrics. A strain (or stress) can be applied either in the direction of the polar axis, or at right angles to it, resulting in two configurations commonly used for piezoelectric generators, termed '*33*' and '*31*' modes as shown schematically in Fig. 4.3. Other configurations are possible (such as shear, '*15*' mode), and the situ-

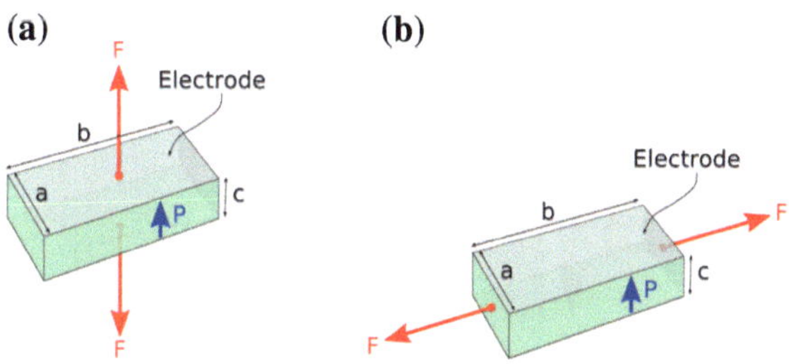

Fig. 4.3 33 (**a**) and 31 (**b**) piezoelectric stress-driven generator configurations. ***F*** is the applied force, ***P*** is polarisation direction, *a*, *b* and *c* are linear sizes of piezoelectric elements. Maximum energy per cycle is $d_{33}g_{33}F^2c/(ab)$ for the 33 generator or $d_{31}g_{31}F^2/a$ for the 31 generator (reproduced from paper by Bowen et al. [24], with permission from The Royal Society of Chemistry)

ation is more complex for materials with lower symmetry, but the '*33*' and '*31*' configurations apply to most practical piezoelectric energy harvesters.

The performance of an energy-harvesting material is directly related to the piezoelectric coefficients and electromechanical properties of the material (see Chaps. 2 and 3), however the applied stress or strain is also an important factor. Efficient coupling between the mechanical source and the piezoelectric material is therefore an important factor in determining the energy-harvesting performance. The energy output also depends on the ability of the piezoelectric material to sustain and withstand the applied force or to repeatedly undergo a recoverable strain without degradation or fatigue of the mechanical or electrical properties. This is particularly important for kinematic energy harvesters, for example a material attached to a deforming tyre wall and subjected to large strains. In many cases it is these limits in the strength and elasticity of the materials that may be the dominant factors in the energy-harvesting performance rather than simply the piezoelectric coefficients.

Piezoelectric inertial vibration harvesters exploit the same piezoelectric properties, but the strain in the piezoelectric material is created by the inertia of a suspended mass undergoing acceleration, rather than being directly deformed by the source. There are many ways of achieving this coupling, but the most common approach is to use a simple piezoelectric cantilever configuration, as shown in Fig. 4.4.

The piezoelectric cantilever is clamped at one end (termed the cantilever 'root') to the vibration source. A mass is fixed to the other end, and when the source accelerates, the inertia of the tip mass induces bending in the piezoelectric cantilever. The magnitude of the tip mass can be used to tune the resonant frequency of the harvester. Bending of the piezoelectric element in Fig. 4.4 leads to the creation of equal and opposite strains on the inside and outside of the bend (the upper and lower surfaces of the cantilever). For a piezoelectric this will lead to the cancellation of electrical charge, so that no net current is generated. To be effective as a generator, it is necessary to move the piezoelectric layer away from the neutral axis and this is often achieved either by fixing the piezoelectric material to a

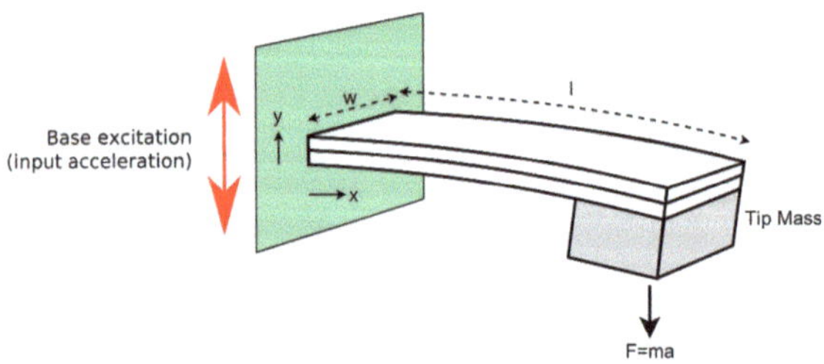

Fig. 4.4 Schematic diagram of the piezoelectric cantilever vibration harvester. The external force ***F*** is applied (reproduced from paper by Bowen et al. [24], with permission from The Royal Society of Chemistry)

non-piezoelectric elastic layer (a 'unimorph'), or by joining two piezoelectric layers that are poled in opposite directions (a 'bimorph') [24].

4.1.2 *Resonant and Non-resonant Performance*

As described in Sect. 4.1.1, the inertial harvesting devices are usually operated in a frequency range at or close to resonance, where the amplitude of the tip oscillation is limited by the losses from the mechanical system resulting from the energy harvested along with internal and external losses due to friction, internal electrical losses and air damping. As a result, the most effective energy harvester does not necessarily employ material with the highest piezoelectric coefficients. Poled PZT FC is a widely used piezoelectric material and is obtainable in a range of compositions from "hard" ferroelectric materials which have low losses but small piezoelectric coupling, through to "soft" ferroelectric materials with much higher piezoelectric coupling, but also much higher losses. Table 1.2 shows the electromechanical properties of these FC materials along with a range of other well-known FCs for comparison. It has been shown that the hard materials, which have smaller piezoelectric coefficients than soft materials, can produce larger power output [29]. However, the best FC material is not immediately obvious, and its performance depends on the magnitude of the electrical power harvested compared to other sources of loss, i.e., the efficiency, and non-harvested losses dominate for many systems. This demonstrates the importance of efficiency, not only in controlling the loading of the source, but also in optimising power output and material selection.

Piezoelectric energy harvesters do not operate under the same thermodynamic constraints as thermal converters and, in theory, the efficiency of conversion could reach 100 % [30]. In practice the losses are usually significantly larger than the energy converted, and typical efficiencies in the order of 20 % [31]. In some cases the mechanical source is not an infinite supply, and the harvesting of energy damps the vibration producing it. In these cases, one can only hope to extract at most the power available from the source and this is best done with a high efficiency, low loss harvester.

There are a range of loss mechanisms which can include air friction and the influence of the clamping arrangement. Internal losses due to ferroelastic hysteresis and inelastic behaviour at joints and interfaces can also contribute. Electrical losses can occur internally before any energy is transferred to the load and are due to capacitive loading of regions of the piezoelectric element that are not being strained significantly; charge flows from the high strain regions to the low strain regions resulting in loss. The tip of the cantilever is only subjected to a small strain, therefore concentrating the piezoelectric material towards the root provides the most effective use of material. It has been shown that for a rectangular cantilever, a piezoelectric coverage of exactly 2/3 of the beam area produced the maximum power output [26]. Positioning the piezoelectric material under the clamp can also significantly increase losses. It should therefore be avoided, although thinning of the structure at the root can reduce stiffness and cause the maximum strain to be developed away from the piezoelectric region and thus reducing overall effectiveness. Optimisation methodologies can be used to maximise harvester performance and will be discussed later in this chapter (see Sect. 4.3).

4.1.3 Linear and Non-linear Performance

One of the limitations of a resonant-based harvesting device is that the power output decreases rapidly away from the resonance frequency [32]. This means that they are only effective in a situation where there is a large component of vibration amplitude at or near to the resonant frequency; in many cases the vibration source consists of multiple or time-varying frequencies. To overcome this limitation a number of strategies have been pursued to increase the bandwidth of energy harvesting devices [33, 34]; this include the use of tuneable resonators, multi-frequency arrays, and non-linear oscillators. Non-linearity is often introduced by engineering two stable states in the device (bistability). Above a specific amplitude the system can switch between the two states in a highly non-linear, non-resonant and chaotic manner. This lack of a well-defined resonant frequency means that the device is effective across a wider frequency range. A recent study of the use of a non-linear piezoelectric harvester to power a heart pacemaker [35] showed that the device was effective from 7 beats per minute to 700 beats per minute. The pacemaker harvester used opposing permanent magnets to create the bistability. Recent reviews on bistable harvesting [36, 37] have classified the potential methods to induce bistability, such as employing magnetic attraction or repulsion on cantilever structures and imparting mechanical bistability into a piezoelectric structure, for example, by engineering asymmetric composite laminates supporting the piezoelectric [38], as shown in Fig. 4.5.

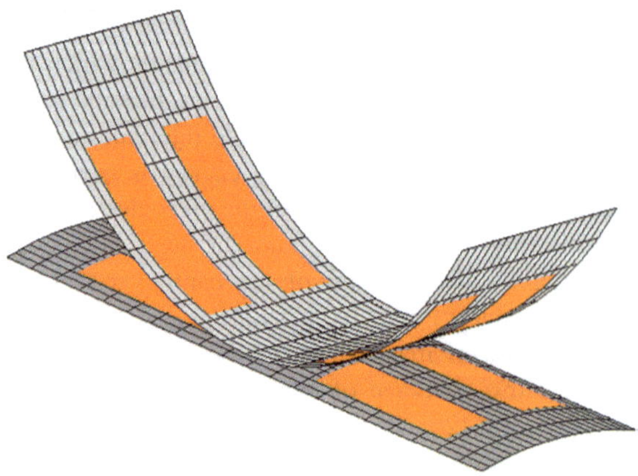

Fig. 4.5 Two stable states of a bistable $[0^P/0/90/90^P]_T$ laminate. *Orange* regions are locations of piezoelectric material (reproduced from paper by Betts et al. [38], with permission from the American Institute of Physics)

An important advantage of piezoelectric materials for energy harvesting is their scalability to a small device size. Integrating the piezoelectric element with silicon electronics using MEMS (Micro-Electro-Mechanical System) fabrication methods offers the promise of low cost high volume self-powered electronic devices and much work has gone into developing devices and processes to make this possible [28]. High-performance piezoelectric materials such as PZT-type FCs can present problems with regard to process compatibility, but significant progress has been made in integrating CMOS (i.e., complementary metal–oxide–semiconductor) compatible materials such as AlN. Although the piezoelectric coefficients d_{ij} of AlN are lower than those of PZT (see data in Table 4.1), Elfrink et al. [53] demonstrated that the piezoelectric coupling compared more favourably due to its low dielectric constant, high squared figure of merit $(Q_{33})^2 = d_{33}g_{33}$, and it is also a lead-free alternative to PZT-based FCs [54]. Microgen recently announced commercial scale production of AlN-based piezoelectric MEMS energy harvesters [55]. Effective micro-scale development requires measurement techniques for measuring the piezoelectric performance at the scale of interest, so recent work has developed MEMS metrology devices to measure the piezoelectric performance at the micro-scale, potentially in situ or in production [56].

In addition to efforts to develop small-scale piezoelectric harvesters, there is also strong research interest in developing nanostructured materials to provide novel energy-harvesting devices and routes to production [5, 9]. ZnO is a piezoelectric material that can be grown as nano-rods on a large scale and significant improvements in the energy-harvesting performance have been reported in work [6, 12, 57, 58]. Power densities in the region of 0.2 $W \cdot cm^{-3}$ have been reported [59] based on measurements of peak short-circuit current and open-circuit voltage for an impulsive mechanical excitation. Recent work on measurement techniques for nano-generators has shown that, in common with the inertial vibration harvesters described above, the output power is dependent on the electrical load, and that the power delivered to the load, averaged over multiple loading cycles, will be smaller than the instantaneous peak power [60] and dependent on the source of excitation. The ability to produce a material that is both functional and manufactured through low cost and energy efficient processes is valuable when considering the development of a new energy-harvesting system. When compared with a number of piezoelectric materials such as poled PZT FC, there are clear environmental benefits. However, since ZnO is non-ferroelectric, its piezoelectric coefficients (d_{33}, d_{31} and d_{15}), as those of AlN and GaN, are relatively small in comparison to d_{ij} of the PZT-type FC (see Tables 4.1, 2.1 and 1.2). Computational investigations of size effects in ZnO nanowires have shown that piezoelectric properties may be enhanced as the diameter of the nano-rods is reduced below around 1.5 nm [61], although current growth methods produce nano-rods with diameters in the range 10–100 nm [44]. Experiments have also shown a possible increase in the piezoelectric effect in GaN nanowires compared to bulk material [40] while nano-scale ferroelectrics have been recently reviewed by Varghese et al. [62].

The operation of piezoelectric materials to enhance energy conversion from a variety of natural sources relies on the development of the piezoelectric potential

Table 4.1 Comparison of room-temperature electromechanical properties of some piezoelectric materials

	GaN	AlN	ZnO	$BaTiO_3$ FC	PZT-4 ('hard' FC)	PZT-5H ('soft' FC)	Domain-engineered PMN–0.33PT SC	$LiNbO_3$ SC	Poled PVDF
d_{33}, pC/N	3.7 [39] 13.2 (NW) [40]	5 [39]	12.4 [43] 14.3–26.7 (nanobelt) [44]	149 [45]	289 [45]	593 [45]	2820 [48]	6 [49]	–33 [51]
d_{31}, pC/N	–1.9 [39] –9.4 (NW) [40]	–2 [39]	–5.0 [43]	–58 [45]	–123 [45]	–274 [45]	–1330 [48]	–1.0 [49]	21 [51]
d_{15}, pC/N	3.1 [41]	3.6 [41]	–8.3 [43]	242 [45]	495 [45]	741 [45]	146 [48]	69 [50]	–27 [51]
k_{33}	–	0.23 [42]	0.48 [43]	0.49 [46]	0.7 [47]	0.75 [47]	0.94 [48]	0.23 [50]	0.19 [52]

Notes
1. Experimental data on electromechanical properties of numerous FCs and relaxor-ferroelectric SCs are given in Tables 1.1, 1.2, 2.4, and 2.5
2. Data for GaN nanowires are marked '(NW)'

through the development of a strain in the material. Since piezoelectrics, and indeed ferroelectrics, can be treated as semiconductors that can sustain a crystal dipole, there is an intimate relationship between the semiconductor properties of the material and any device behaviour resulting from a strain. This relationship has long been understood for piezoelectric materials with early reports correctly identifying barrier height changes in III–V semiconductor structures [63] and strain induced piezoelectric effects in GaN optoelectronic devices [64]. These reports have led to a variety of devices being produced that exploit these relationships for a range of piezoelectric materials. The development of such devices has rapidly come to maturity, and there are now a large number of applications where the piezoelectric, or ferroelectric nature, of a device is harnessed to generate energy under a controlled environment [60]. There is also now growing evidence that ferroelectric materials, such as $BaTiO_3$, $LiNbO_3$ and PZT are true semiconductors, and we marked the semiconductor feature of ferroelectrics in the diagram (see Fig. 4.1, left part). For example, undoped PZT FC is a wide-band gap semiconductor with a band gap of between 2.6 eV and 3.5 eV [65]. The PZT FC material also exhibits p-type electric conductivity due to the presence of low-valence impurities substituting for higher-valence Pb ions. This causes A-site (Pb ion) vacancies to act as electron acceptors, leading to the production of holes in a FC sample [66]. The behaviour is modified due to the non-centrosymmetric crystal structure, and this can be used to enhance a number of interesting device and materials performance parameters, such as photo-voltaic performance or photochemical yield.

4.2 Physical Characteristics of Piezoelectric Materials for Energy Harvesting

4.2.1 *High-Temperature Harvesting*

A considerable amount of research has concentrated on vibration harvesting at ambient temperatures since one motivation is to power low-power electronic devices and wireless systems. However there are a number of applications, such as power, transport or oil/gas/space exploration, where there is a need to operate at higher temperatures. As an example, temperatures up to 600 °C are widely encountered in engines of different types and industrial processes.

In terms of the piezoelectric material for high-temperature , many ferroelectric materials are characterised by a Curie temperature $T_C < 600$ °C. For example, PZT-based ferroelectrics have a Curie temperatures $T_C < 400$ °C and gradual reduction in power with temperature up to 150 °C has been reported for soft PZT harvesters [67]. Comyn et al. [68] have recently processed $BiFeO_3$-based polycrystalline ceramics with T_C up to 650 °C and $Bi_4Ti_3O_{12}$, another ferroelectric, has been shown to be stable up to 500 °C and has been considered as a potential material for use at the ambient temperature of Venus (460 °C) [69].

In addition to ferroelectrics, wide-band gap semiconductor and piezoelectric and piezoelectric materials with wurtzite structures are of interest. GaN is a potential piezoelectric material that exhibits the semiconductor and piezoelectric properties that is advantageous for the realisation of high-temperature harvesting. While the piezoelectric coefficients d_{ij} are not as high as in ferroelectrics (compare GaN with the PZTs in Table 4.1), GaN nanowires have demonstrated high piezoelectric coefficients [40], and piezoelectric sensors based on GaN have been reported [70]. In Table 4.1 we show piezoelectric data for GaN nanowires and bulk SC samples. Due to their wide band gap, these materials are expected to operate in a broad temperature range and retain low electric conductivity, and being semiconductors have the potential to integrate with device electronics associated with the energy harvester. In terms of device electronics, the relatively narrow band gap of silicon results in device functionality being degraded at temperatures in the region of about 350–400 °C since the intrinsic densities of electrons and holes become significant compared with doping densities. The use of wideband gap materials, such as GaN or SiC, is one possible solution for energy harvesting in hostile environments [71].

Another potential high-temperature material is AlN. Like GaN, AlN has a wurtzite crystal structure, does not exhibit a phase transition on heating and has a melting point of 2000 °C [72]. As is known from work [73], *c*-axis orientated thin films of AlN have been used in high-temperature piezoelectric transducers. Piezoelectric activity in AlN is observed even at temperatures as high as 1150 °C, and the material also has low electric conductivity owing to its large band gap. AlN can be used at low oxygen partial pressures, an advantage if the transducer must operate under reducing conditions to prevent oxidation of packaging.

Ferroelectric $LiNbO_3$ is another option for high-temperature piezoelectric operation and has been considered for high-temperature actuation and sensing in harsh environments for applications such as ultrasonic drills, corers, and rock abrasion tools [74]. Under shear conditions $LiNbO_3$ SC shows relatively large piezoelectric activity (see d_{15} in Table 4.1) and electromechanical coupling factors, a pre-requisite for efficient energy conversion, and a very high Curie temperature (T_C = 1142–1210 °C) [75, 76]. By using high-purity $LiNbO_3$ SCs, transducers operating at temperatures up to 1000 °C have been reported with no significant oxygen loss or resistance change over 600 °C [63], but there is less work on energy harvesting using this material [77]. Bedekar et al. [78] have shown that $YCa_4O(BO_3)_3$ and $La_3Ga_5SiO_{14}$ exhibit stable piezoelectric and dielectric properties up to 1000 °C. $GdCa_4O(BO_3)_3$ piezoelectric SCs have also been considered for ultra-high temperature (>1000 °C) applications [79]. Zhang et al. [80] have provided an excellent overview on piezoelectric sensor materials for high-temperature applications. While there is evidence of research on using piezoelectric materials for high-temperature transducers, such as sensors, there is less work specifically on harvesting and the associated circuits and storage challenges under extreme conditions.

4.2.2 Compliant Piezoelectrics

Many of the piezoelectric materials described in Sects. 4.1 and 4.2 (e.g. PZT, GaN, ZnO, and $LiNbO_3$) are inorganic and inherently hard, high stiffness and brittle with the potential to fracture at low tensile strains [81]. Polymeric materials that exhibit piezoelectric behaviour are of independent interest for energy harvesting due to their flexibility, toughness, low density, biocompatibility, and low-cost characteristics. Example applications include wearable or implanted devices [82] where the polymer material is subjected to bending or stretching by limb motion or lung and cavity expansion during respiration [11]. The most common piezoelectric polymer is PVDF whose piezoelectric behaviour originates from orientated molecular dipoles, which are formed by a combination of mechanical deformation and electrical poling of ferroelectric β-phase PVDF [81]. The material has a failure strain of 2 % or higher [11]; however the piezoelectric coefficients d_{ij} and ECFs k_{ij} are relatively low in comparison to the parameters of the aforementioned inorganic piezoelectrics (see Table 4.1). Example applications include harvesting footsteps (using foils) [83], respiration (using microbelts) [84], rainfall and wind [85]. As with the inorganic materials, methods to manufacture PVDF at the nano-scale have been developed by employing electro-spinning using a needle [11] or disc [81] to form nano-fibre webs. The formation of β-phase PVDF for optimum piezoelectric properties can be promoted since the electro-spinning process can be undertaken under high electric field and mechanical stress.

Ferroelectret polymers have been explored for vibration harvesting to a lesser extent. In this case their piezoelectric properties originate from internally charged voids [86, 87]. Their potential advantage over PVDF are the higher piezoelectric coefficients (e.g. d_{33} > 200 pC/N) and greater elastic compliance [88] (e.g. $s_{11} \approx 1100\ \text{pPa}^{-1}$ [89]); although they have low electromechanical coupling factors (e.g. k_{33} < 0.1 [90]), and the output voltage was observed to degrade at lower temperatures compared to PVDF [86]. Other attempts to create flexible piezoelectric structures is the use of fine-scale 'wavy' PZT ribbons on flexible rubber substrates to allow stretching and flexing modes for harvesting biomechanical energy [91, 92]. In this way the high piezoelectric activity of the PZT FC material can be exploited, especially when enhanced by operating in a buckling mode. Such piezoelectric 'nano-ribbons' have also been used to form a biocompatible interface with cells to act as sensors, with the potential to act a harvester of biomechanical sources [93]. We add that bio-piezoelectric devices have also been considered based on genetically modified bacterial viruses (M13 phages) with an aligned protein coat structure to form the necessary electrical dipole [94].

In summary, whilst piezoelectric transducers have been studied for many years both as sensors and actuators, it is only recently that significant attention has been devoted to their use as an energy source. Whilst the fundamental principles of piezoelectric coupling of electrical and mechanical energies are unchanged, there are many complexities associated with their application to energy production that have only recently been addressed. As the technology develops, new opportunities will arise for new materials, techniques and innovations. As the technology moves towards production scale-up and wider market penetration, this knowledge will need to be transferred to industrial standards for device performance and reliability.

4.3 Optimisation for Piezo-Based Harvesting

Energy-harvesting devices and systems are complex multi-physics systems that require advanced methodologies to maximise their performance. Section 4.3 examines the approaches that optimise these complex systems via mathematical programming or population-based optimisation techniques such as the genetic algorithm. The majority of the literature to date has considered the conversion of mechanical vibrations into electrical energy conversion using piezoelectric materials [95, 96]. Due to simplicity and the existence of the well-understood models, cantilever beams, as described in Sect. 4.1, continue to be the system of choice for optimisation studies, although in recent years, researchers have begun to optimise two-dimensional plates. Design optimisation for harvesting elements comprises of a coupled system of three elements. The first element is the dynamic response of the harvesting structure, which is a mature field. The second element is the electrical circuit that is necessary to condition the voltage and charge that is generated, this is an area where there has been active development [97–101]. The final element is the coupled electro-mechanics of the system, which represents the key step in energy harvesting and poses a challenging multi-physics problem to the optimisation community. Some of the published research has considered optimisation of material properties, e.g. optimisation of the microscopic crystallite configuration to maximise the electromechanical coupling in specific ferroelectric materials [102]. However, the majority of research has taken a structural optimisation approach to piezoelectric energy-harvesting systems at the device level. This section examines these research efforts and the methods employed.

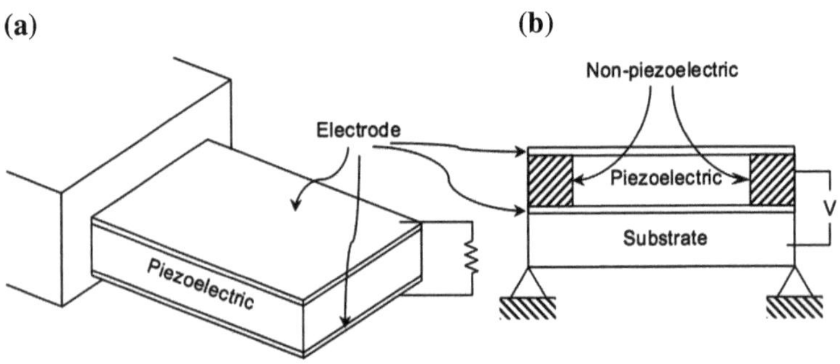

Fig. 4.6 Common piezoelectric energy-harvesting configurations: cantilevered beam (**a**) and cross-section of a plate (**b**) (reproduced from paper by Bowen et al. [24], with permission from The Royal Society of Chemistry)

The most common configurations for optimisation studies are cantilevered beam and plates, examples of which are shown in Fig. 4.6. Studies have shown that the lay-out, position and configuration of the piezoelectric material can have a significant influence on the energy-harvesting performance of a device [103–105]. Dietl and Garcia [104] used a combination of the constrained pattern search algorithm and gradient search method to optimise the width of a bimorph cantilever beam with a tip mass to maximise the voltage generation over time. The first two modes of vibration were included for the optimisation study and it was found that the optimum configuration of the beam tapered down from the root and then widened again near tip. Other researchers also found higher and more uniform strain areas could be achieved by developing a trapezoidal tapered beam along the beam span [105] and through-thickness [106], thereby leading to an increase in both the specific output power per unit volume. In addition, Goldschmidtboeing and Woias [107] obtained an optimum harvester configuration for a unimorph cantilever beam device of plan-form geometry; this was a triangular geometry for a single excitation mode. Here they defined the optimum performance to be the device efficiency, which was characterised by the output power and the maximum tolerable amplitude, taking into account the stress homogeneity. Interestingly, they found the plan-form shape to have little impact on the overall efficiency but is highly sensitive to stress which, in turn, influences the tolerable vibration amplitude. This supports the observation of Wang [108] in that under a static load the plan-form structure has little influence on the efficiency of the electrical energy conversion, but a trapezoidal cross-section enhances the output voltage.

In contrast to the effort on the optimisation of linear energy harvesters presented in this section, Betts et al. [38] optimised a non-linear bistable piezo-composite energy harvester, described in Sect. 4.1.3, using sequential quadratic programming. The design space was highly nonlinear and multimodal however, it was possible to consistently find all local and global optima by employing multiple random starting solutions [38]. As with the above studies, the dimensional parameters of the rectangular plate geometry were optimised to maximise the energy output characterised by the maximum strain. Due to the nonlinear nature of the bistable structure, the strain is large and the power output can be as much as an order of magnitude greater than a linear harvester with an added benefit of harvesting appreciable energy over a broad spectrum of excitation frequencies [109].

The investigations of the shape of the structure for energy-harvesting device have so far been limited in that the majority of the literature either constructs a simplified analytical model or a reduced order model and then conducts the optimisation analytically or study a small set of geometries. Using linear elasticity, much of the understanding of the optimum energy-harvesting performance relates to the precise position of the piezoelectric material; this is usually in high strain areas of the device and in these cases the lowest bending mode is most beneficial. As such work only considers quadrilateral and triangular geometries, the design space is inherently limited and the understanding of the optimum design and potential of energy harvesting are also limited. Park et al. [110] opened up the design space by applying shape optimisation to the beam planform. Since there was

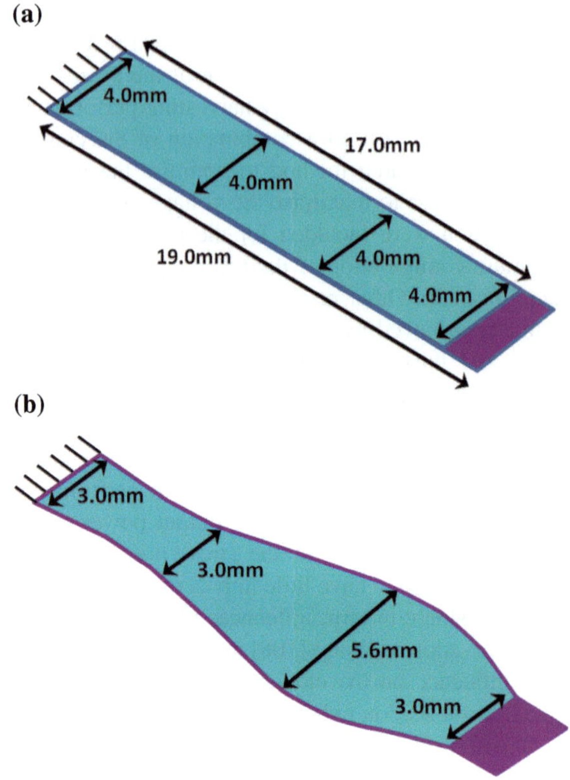

Fig. 4.7 **a** Initial and **b** optimum bimorph beam shape for piezoelectric cantilever beam for maximum power by rotary motion. The bimorph is in parallel connection clamped at upper left hand side end of the beam. *Blue* is the piezoelectric and purple is the substrate where the tip region is deflected. Length dimensions are fixed. Adapted from work by Park et al. [110] (reproduced from paper by Bowen et al. [24], with permission from The Royal Society of Chemistry)

limited design space in their application domain, they specify a maximum length of the cantilever beam and they allow an arbitrary width variation to maximise the output power of the energy harvester, as shown in Fig. 4.7. The optimum solution (Fig. 4.7b) was demonstrated to achieve 37 % improvement compared with the rectangular plan-form of the same volume (Fig. 4.7a).

Researchers consistently agree that consideration of the electromechanical coupling of the harvester is important [111]. Work on addressing this complex multi-physics problem employs stochastic optimisation, which does not require an analytical model or gradient sensitivities. Gonsalez et al. combined a genetic algorithm (GA) [112], which is a heuristic search algorithm based on natural selection to evolve a population of potential solutions with a reduced order model to maximise the power output of a piezoelectric-substrate beam combination. This approach enabled optimisation of the piezoelectric and substrate thickness, mechanical loss factor and electrical impedance [112]. Benkhelifa et al. used a well-established genetic algorithm for multiple objectives, MOGA-II, to maximise the harvested power and voltage output whilst minimising the size of a bimorph piezoelectric beam when it was subjected to a single excitation frequency [113]. Bourisli and Al-Ajmi optimised a unimorph cantilever beam to maximise the conversion of mechanical energy to electrical energy for the first three vibration modes of the device using a genetic algorithm [114]. Bourisli and Al-Ajmi [114] examined the optimum piezoelectric coverage pattern for different substrate materials, namely brass, steel and aluminium. The study revealed that the optimum designs are not influenced by the choice of the substrate materials, and the optimum location of the piezoelectric material coincided with the regions of maximum strain for each vibration mode. We also note the optimisation study of Hadas et al. [115], who applied a Self-Organisation Migrating Algorithm (SOMA) which mimics the behaviour of wild animal groups; although their application domain was electromagnetic vibration energy harvesting. For their numerical studies of multi-objective optimisation, SOMA was considered superior to GA although they are both able to find the optimum solutions. Gurav et al. studied the maximisation of the power output of small-scale MEMS-based energy harvesters [116]. In order to address the manufacturing challenges and control of the material properties and microstructure, uncertainty-based design optimisation was applied to determine an optimum combination of geometric variables.

Understanding that the shape has a significant effect on an energy harvester, the approaches so far consider only a small number of geometrical variables and have explored a relatively small design space, i.e. mainly studying well-defined geometries such as rectangles and trapezoids. In addition, the complex multi-physics dynamics of energy harvesting is not well understood and an 'intuitive' design may not be an optimum. In order to explore a greater design space to include unintuitive designs, researchers have developed topology optimisation for linear energy harvesters. Topology optimisation is a class of structural optimisation that provides the most creative solution independent of the preconceived or initial design. In the last decades, topology optimisation is gaining popularity in many different domains of physics [117]. Two categories of approaches have emerged over the past three decades. The traditional elemental approach is to formulate the design problem as a material distribution problem where each small unit or element of material is considered to be a design variable which can take either 1 (material) or 0 (void). The design space therefore becomes the distribution of material and voids, which represents the topology or the general layout solution [118]. The most popular methodology in this category of approaches is termed 'Solid Isotropic Material with Penalisation' (SIMP). The alternative approach is commonly referred to as the *Level Set Method* which represents the structural boundaries as a set of implicit signed distance functions and the boundaries are moved to minimise the objective function, thus producing the topologically optimum solution [119, 120].

Topology optimisation has been applied to piezoelectric energy-harvesting systems. The more common approach is an extension of SIMP where the key approach is to relax the binary design variables to a continuous variable between 0 and 1 ($0 < x \leq 1$) then penalise the infeasible solutions using a power law, where $x \in \{0, 1\}$. Extending this to the problem of piezoelectrics, the power law is applied to elastic stiffness, piezoelectric and dielectric parameters [121]. Researchers found

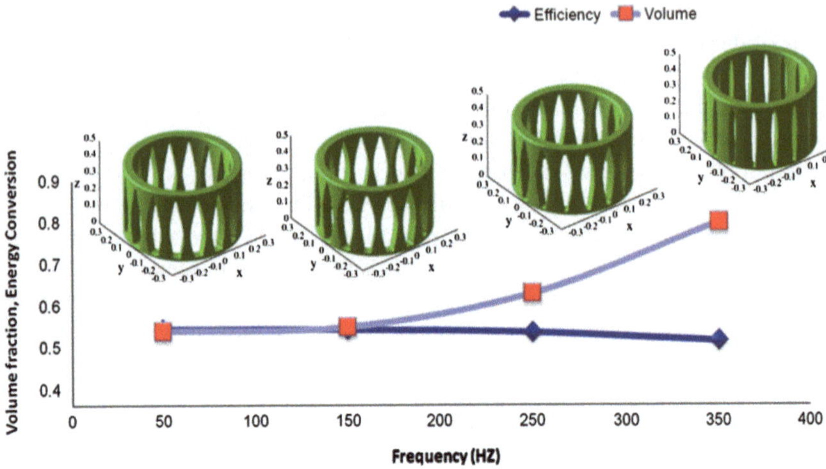

Fig. 4.8 Optimum topology for piezoelectric energy harvester operating in the '33' mode, using the level set method (reprinted from paper by Chen et al. [124], with permission from Elsevier)

that the choice of the three exponents can lead to different solutions or even non-convergence, thus they need to be carefully selected [122, 123]. The alternative level set method avoids the challenges associated with using the power law [124], and both the '31' and '33' operation modes, shown in Fig. 4.3, have been considered using this approach.

Chen et al. [124] applied the level set method to optimise a cylindrical energy harvester using two materials that operated in the '33' mode, see Fig. 4.8. Optimisation for multiple materials was achieved by a '*Reconciled Level Set*' method [125]. Sun and Kim [126] also optimised two materials in a magneto-electro-elastic laminate composite. In their study the thickness of each material was optimised using a micro-mechanics based model under the influence of a static load. They found that the SIMP-type material interpolation model was unable to converge to a solution with distinct phase states for this complex multi-physics problem. By far the more common configurations considered in the topology optimisation literature remain to be simple flat cantilevered beams and plates as in Fig. 4.6 [121–123, 127–129], albeit there are variations, e.g. with or without substrates, with or without tip mass. Rupp et al. [127] optimised the electrical circuit parameters simultaneously with the general layout. Their numerical studies showed that the simultaneous optimisation of both structural topology and circuit did not fundamentally change the topological designs, but did influence the optimal resistance of the circuit.

For energy-harvesting devices designed for quasi-static applications or at an excitation frequency much lower than the resonance frequency, the optimisation objective tends to be to maximise the ECF k_{ij} [126, 128–130] since it characterises an efficiency of conversion between mechanical and electrical energies in a

Fig. 4.9 Optimum cantilever beam topology of piezoelectric material under excitation frequencies 10–400 Hz. The beam is configured as in Fig. 4.7a, clamped on upper left side. *Black* represents the material is 1 and white is void, 0. Grey is an intermediate density between 0 and 1 (reprinted from paper by Lin et al. [131], with permission from Springer)

piezoelectric medium. In Chap. 2 we analysed examples of ECFs of numerous piezoelectric materials at various oscillation modes.

In a case of dynamic applications, the harvested power for the given vibration environment is maximised [127, 130–132]. Wein et al. [132] added a stress constraint to their optimisation to control the peak stress in a piezoelectric and substrate composite system using linear elasticity so that the device can sustain an applied force or to repeatedly undergo a recoverable strain. It is interesting to note that Chen et al. [124] optimised a dynamic system but used a mean steady-state energy conversion efficiency similar to the ECF.

Most dynamic optimisation studies considered a single frequency environment and the structural layout was optimised to 'tune' its resonant modes to the excitation frequency. Lin et al. [131] optimised a cantilevered beam energy harvester for broadband random vibration (Fig. 4.9). Comparing solutions for objective functions and for a broadband environment, one can state that their topological designs are fundamentally different. Thus, more research is needed in dynamics and particularly for broadband and random ambient vibration conditions. We also note that there exists very little research beyond the cantilever beam and plate configurations and optimisation of more recently developed harvesters such as piezoelectric nano-generators would be a fruitful avenue for future research and discussions.

4.4 Conclusion

Energy harvesting remains a topic of intense interest, and this chapter provides a timely overview of the variety of energy-harvesting mechanisms employed by piezoelectric and related materials due to their electromechanical coupling.

Piezoelectric materials are solid-state dielectric materials that provide effective conversion between electrical and mechanical energy and can be manufactured at small scales and integrated into micro-scale devices or even electronic circuits. There are a number of potential materials and device configurations and properties and loss mechanism need also be considered, along with the potential scale of the system (from 10^{-2} to 10^{-9} m). One of the main limitations of resonant-based devices is that their power output decreases rapidly away from the resonant frequency and non-linearity can be introduced to enable more 'broadband' harvesting. As vibration harvesting matures it is likely that they will need to be deployed in more hostile environments. Promising candidate materials for such applications include high Curie temperature ferroelectrics and wideband gap semiconductor materials, although the associated circuits and storage challenges under extreme conditions must also be met. In applications, where a high strain is required, more compliant systems based on polymers or composite systems are being considered.

The electromechanical coupling in piezoelectric-based devices and the complex dynamic response harvester devices mean that optimisation of the harvesting device remains a complex task. In many cases simplified analytical models or reduced order models are used to optimise analytically or study a small set of geometries. A variety of approaches are however available to maximise performance in terms of efficiency or total power; this can include methods to optimise an 'initial' design or topology optimisation which has the potential to develop a creative solution independent of the preconceived or initial design. To date most optimisation studies have concentrated on mechanical energy harvesting, although there is potential to apply these approaches to other systems, such as pyroelectric harvesting or even combined harvesting systems.

In summary, piezoelectrics, pyroelectrics and ferroelectrics (Fig. 4.1) represent main groups of active dielectrics and can be regarded as important materials for energy-harvesting applications not any due to their efficiency, ease of use, scale, integration with electronics but also because of their versatility and the variety of modes that they can be deployed.

References

1. Perpetuum Ltd, the World Leader in Vibration Harvester Powered Wireless Sensing Systems. http://www.perpetuum.com/rail/
2. Mane P, Xie J, Leang KK, Mossi K (2011) Cyclic energy harvesting from pyroelectric materials. IEEE Trans Ultrason Ferroelectr Freq Control 58:10–17
3. Guan MJ, Liao WH (2008) Characteristics of energy storage devices in piezoelectric energy harvesting systems. J Intell Mater Syst Struct 19:671–680
4. Anton SR, Erturk A, Inman DJ (2010) Multifunctional self-charging structures using piezoceramics and thin-film batteries. Smart Mater Struct 19:115021
5. Wang ZL, Wu W (2012) Nanotechnology enabled energy harvesting for self-powered micro/nanosystems. Angew Chem 51:2–24
6. Wang ZL (2012) From nanogenerators to piezotronics−A decade long study of ZnO nanostrucutres. MRS Bull 37:814–827

7. Wang ZL (2012) Progress in piezotronics and piezo-phototronics. Adv Mater 24:4632–4646
8. Wang ZL, Yang R, Zhou J, Qin Y, Xu C, Hu Y, Xu S (2010) Lateral nanowire/nanobelt based nanogenerators, piezotronics and piezo-phototronics. Mater Sci Eng, R 70:320–329
9. Wang ZL, Zhu G, Yang Y, Wang S, Pan C (2012) Progress in nanogenerators for portable electronics. Mater Today 15:532–543
10. Wang X (2012) Piezoelectric nanogenerators−Harvesting ambient mechanical energy at the nanometer scale. Nano Energy 1:13–24
11. Qi Y, McAlpine MC (2010) Nanotechnology-enabled flexible and biocompatible energy harvesting. Energy Environ Sci 3:1275–1285
12. Wang ZL (2009) ZnO nanowire and nanobelt platform for nanotechnology. Mater Sci Eng, R 64(3–4):33–71
13. Hiralal P, Unalan H, Amaratunga GAJ (2012) Nanowires for energy generation. Nanotechnol 23:194002
14. Radousky HB, Liang H (2012) Energy harvesting: an integrated view of materials, devices and applications. Nanotechnol 23:502001
15. Cook-Chennault KA, Thambi N, Sastry AM (2008) Powering MEMS portable devices−A review of non-regenerative and regenerative power supply systems with special emphasis on piezoelectric energy harvesting systems. Smart Mater Struct 17:043001
16. Kim HS, Kim JH, Kim J (2011) A review of piezoelectric energy harvesting based on vibration. Int J Precis Eng Manuf 12:1129–1141
17. Sodano HA, Inman DJ, Park G (2004) A review of power harvesting from vibration using piezoelectric materials. Shock Vib Digest 36:197–205
18. Nechibvute A, Chawanda A, Luhanga P (2012) Piezoelectric energy harvesting devices: An alternative energy. Smart Mater Res 2012:853481
19. Mitcheson PD, Yeatman EM, Rao GK, Holmes AS, Green TC (2008) Energy harvesting from human and machine motion for wireless electronic devices. Proc IEEE 96:1457–1486
20. Kitio Kwuimy CAK, Litak G, Borowiec M, Nataraj C (2012) Performance of a piezoelectric energy harvester driven by air flow. Appl Phys Lett 100:024103
21. Dias JAC, De Marqui C, Erturk A (2013) Hybrid piezoelectric-inductive flow energy harvesting and dimensionless electroaeroelastic analysis for scaling. Appl Phys Lett 102:044101
22. Beeby SP, Tudor MJ, White NM (2006) Energy harvesting vibration sources for microsystems applications. Meas Sci Technol 17:R175–R195
23. Karami MA, Inman DJ (2012) Powering pacemakers from heartbeat vibrations using linear and nonlinear energy harvesters. Appl Phys Lett 100:042901
24. Bowen CR, Kim HA, Weaver PM, Dunn S (2014) Piezoelectric and ferroelectric materials and structures for energy harvesting applications. Energy Environ Sci 7:25–44
25. Donelan JM, Li Q, Naing V, Hoffer JA, Weber DJ, Kuo AD (2008) Biomechanical energy harvesting: generating electricity during walking with minimal user effort. Science 319 (5864):807–810
26. Stewart M, Weaver PM, Cain M (2012) Charge redistribution in piezoelectric energy harvesters. Appl Phys Lett 100:073901
27. Roundy S, Wright P, Rabaey J (2003) A study of low level vibrations as a power source for wireless sensor nodes. Comput Commun 26:1131–1144
28. Kim S, Priya S, Kanno I (2012) Piezoelectric MEMS for energy harvesting. MRS Bull 37:1039–1050
29. Erturk A, Inman DJ (2011) Piezoelectric energy harvesting. Wiley, Chichester
30. Umeda M, Nakamura K, Ueha S (1996) Analysis of the transformation of mechanical impact energy to electric energy using piezoelectric vibrator. Jpn J Appl Phys 35:3267–3273
31. Cho JH, Richards RF, Bahr DF, Richards CD, Anderson MJ (2006) Efficiency of energy conversion by piezoelectrics. Appl Phys Lett 89:104107
32. Erturk A, Inman DJ (2009) An experimentally validated bimorph cantilever model for piezoelectric energy harvesting from base excitations. Smart Mater Struct 18:025009

33. Zhu D, Tudor MJ, Beeby SP (2010) Strategies for increasing the operating frequency range of vibration energy harvesters: a review. Meas Sci Technol 21:022001
34. Tang L, Yang Y, Soh CK (2013) Broadband vibration energy harvesting techniques. In: Elvin N, Erkturk A (eds) Advances in energy harvesting Methods. Springer, New York, pp 17–61
35. Karami MA, Inman DJ (2012) Powering pacemakers from heartbeat vibrations using linear and nonlinear energy harvesters. Appl Phys Lett 100:042901
36. Harne RL, Wang KW (2013) A review of the recent research on vibration energy harvesting via bistable systems. Smart Mater Struct 22:023001
37. Pellegrini SP, Tolou N, Schenk M, Herder JL (2013) Bistable vibration energy harvesters: a review. J Intell Mater Syst Struct 24:1303–1312
38. Betts DN, Kim HA, Bowen CR, Inman DJ (2012) Optimal configurations of bistable piezo-composites for energy harvesting. Appl Phys Lett 100:114104
39. Guy IL, Muensit S, Goldys EM (1999) Extensional piezoelectric coefficients of gallium nitride and aluminum. Appl Phys Lett 75:4133–4135
40. Minary-Jolandan M, Bernal RA, Kuljanishvili I, Parpoil V, Espinosa HD (2012) Individual GaN nanowires exhibit strong piezoelectricity in 3D. Nano Lett 12:970–976
41. Muensit S, Goldys EM, Guy IL (1999) Shear piezoelectric coefficients of gallium nitride and aluminum nitride. Appl Phys Lett 75:3965–3967
42. Tadigadapa S, Mateti K (2009) Piezoelectric MEMS sensors: state-of-the-art and perspectives. Meas Sci Technol 20:092001
43. Crisler DF, Cupal JJ, Moore AR (1968) Dielectric, piezoelectric, and electromechanical coupling constants of zinc oxide crystals. Proc IEEE 56:225–226
44. Espinosa HD, Bernal RA, Minary-Jolandan M (2012) A review of mechanical and electromechanical properties of piezoelectric nanowires. Adv Mater 24:4656–4675
45. Hayward G, Bennett J, Hamilton R (1995) A theoretical study on the influence of some constituent material properties on the behavior of 1–3 connectivity composite transducers. J Acoust Soc Am 98:2187–2196
46. Hong K, Xu H, Konishi H, Li X (2010) Direct water splitting through vibrating piezoelectric microfibers in water. J Phys Chem Lett 1:997–1002
47. Berlincourt D, Krueger HHA Properties of piezoelectricity ceramics. Technical Publication TP-226, Morgan electroceramics, http://www.morgantechnicalceramics.com/sites/default/files/documents/tp226.pdf
48. Zhang R, Jiang B, Cao W (2001) Elastic, piezoelectric, and dielectric properties of multidomain 0.67Pb(Mg1/3Nb2/3)O3–0.33PbTiO3 single crystals. J Appl Phys 90:3471–3475
49. Weis RS, Gaylord TK (1985) Lithium niobate: summary of physical properties and crystal structure. Appl Phys A 37:191–203
50. Turner RC, Fuierer PA, Newnham RE, Shrout TR (1994) Materials for high temperature acoustic and vibration sensors: a review. Appl Acoust 41:299–324
51. Dahiya R, Valle M (2013) Robotic tactile sensing. Technologies and system. Appendix A −fundamentals of piezoelectricity. Springer, Dordrecht
52. Hill CR, Bamber JC, Ter Haar GR (eds) (2004) Physical principles of medical ultrasonics. Wiley, Chichester
53. Elfrink R, Kamel TM, Goedbloed M, Matova S, Hohlfeld D, Van Andel Y, Van Schaijk R (2009) Vibration energy harvesting with aluminum nitride-based piezoelectric devices. J Micromech Microeng 19:094005
54. Andosca R, McDonald TG, Genova V, Rosenberg S, Keating J, Benedixen C, Wu J (2012) Experimental and theoretical studies on MEMS piezoelectric vibrational energy harvesters with mass loading. Sens Actuators A 178:76–87
55. MicroGen Systems, Inc., the global leader in designing and manufacturing nanotechnology/MicroElectroMechanical Systems micro-power and micro-sensor products for wireless sensor and mobile electronics applications http://www.microgensystems.co

56. Wooldridge J, Muniz-Piniella A, Stewart M, Shean TAV, Weaver PM, Cain MG (2013) Vertical comb drive actuator for the measurement of piezoelectric coefficients in small-scale systems. J Micromech Microeng 23:035028
57. Briscoe J, Stewart M, Vopson M, Cain M, Weaver PM, Dunn S (2012) Nanostructured p-n junctions for kinetic-to-electrical energy conversion. Adv Energy Mater 2:1261–1268
58. Briscoe J, Bilotti E, Dunn S (2012) Measured efficiency of a ZnO nanostructured diode piezoelectric energy harvesting device. Appl Phys Lett 101:093902
59. Hu Y, Lin L, Zhang Y, Wang Z (2012) Replacing a battery by a nanogenerator with 20 V output. Adv Mater 24:110–114
60. Briscoe J, Jalali N, Woolliams P, Stewart M, Weaver PM, Cain M, Dunn S (2013) Measurement techniques for piezoelectric nanogenerators. Energy Environ Sci 6:3035–3045
61. Agrawal R, Espinosa HD (2011) Giant piezoelectric size effects in zinc oxide and gallium nitride nanowires. A first principles investigation, Nano Lett 11:786–790
62. Varghese J, Whatmore RW, Holmes JD (2013) Ferroelectric nanoparticles, wires and tubes: synthesis, characterisation and applications. J Mater Chem C 1:2618–2638
63. Yu ET, Dang XZ, Asbeck PM, Lau SS, Sullivan GJ (1999) Spontaneous and piezoelectric polarization effects in III–V nitride heterostructures. J Vac Sci Technol, B 17:1742–1749
64. Flory CA, Hasnain G (2001) Modeling of GaN optoelectronic devices and strain-induced piezoelectric effects. IEEE J Quantum Electron 37:244–253
65. Nagaraj B, Aggarwal S, Song T, Sawhney T, Ramesh R (1999) Leakage current mechanisms in lead-based thin-film ferroelectric capacitors. Phys Rev B 59:16022–16027
66. Jaffe B, Cook W, Jaffe H (1971) Piezoelectric ceramics. Academic Press, London
67. Kim S, Park J, Ahn H, Liu D, Kim D (2011) Temperature effects on output power of piezoelectric vibration energy harvesters. Microelectron J 42:988–991
68. Comyn TP, McBride SP, Bell AJ (2004) Processing and electrical properties of $BiFeO_3$–$PbTiO_3$ ceramics. Mater Lett 58:3844–3846
69. Sherrit S, Bao X, Bar-Cohen Y, Chang Z (2004) Resonance analysis of high-temperature piezoelectric materials for actuation and sensing. Proc SPIE 5387:411–420
70. Le Boulbar ED, Edwards MJ, Vittoz S, Vanko G, Brinkfeldt K, Rufer L, Johander P, Lalinský T, Bowen CR, Allsopp DWE (2013) Effect of bias conditions on pressure sensors based on AlGaN/GaN high electron mobility transistor. Sens Actuators A 194:247–251
71. Kudimi JMR, Mohd-Yasin F, Dimitrijev S (2012) SiC-based piezoelectric energy harvester for extreme environment. Procedia Eng 47:1165–1172
72. Damjanovic D (1998) Materials for high temperature piezoelectric transducers. Curr Opin Solid State Mater Sci 5:469–473
73. Hou R, Hutson D, Kirk KJ, Fu YQ (2012) AlN thin film transducers for high temperature non-destructive testing. J Appl Phys 111:074510
74. Bao X, Bar-Cohen Y, Scott J, Sherrit S, Widholm S, Badescu M, Shrout T, Jones B (2010) Ultrasonic/sonic drill for high temperature application. Proc SPIE 7647:764739
75. Baba A, Searfass CT, Tittmann BR (2010) High temperature ultrasonic transducer up to 1000 °C using lithium niobate single crystal. Appl Phys Lett 97:232901
76. Fraser DB, Warner W (1966) Lithium niobate. A high-temperature piezoelectric transducer. J Appl Phys 37:3853–3854
77. Lewis RWC, Allsopp DWF, Shields P, Šatka A, Yu S, Topolov VYu, Bowen CR (2012) Nano-imprinting of highly ordered mano-pillars of lithium niobate ($LiNbO_3$). Ferroelectrics 429:62–68
78. Bedekar V, Oliver J, Zhang S, Priya S (2009) Comparative study of energy harvesting from high temperature piezoelectric single crystals. Jpn J Appl Phys Pt 1(48):0914061–0914065
79. Zhang S, Frantz E, Xia R, Everson W, Randi J, Snyder DW, Shrout TR (2008) Gadolinium calcium oxyborate piezoelectric single crystals for ultrahigh temperature (> 1000 °C) applications. J Appl Phys 104:084103
80. Zhang S, Yu F (2011) Piezoelectric materials for high temperature sensors. J Am Ceram Soc 94:3153–3170

81. Fang J, Niu H, Wang H, Wang X, Lin T (2013) Enhanced mechanical energy harvesting using needleless electrospun poly(vinylidene fluoride) nanofibre webs. Energy Environ Sci 6:2196–2202
82. Chang J, Dommer M, Chang C, Lin L (2012) Piezoelectric nanofibres for energy scavenging applications. Nano Energy 1:356–371
83. Kymissis J, Kendall C, Paradiso J, Gershenfeld N (1998) Parasitic power harvesting in shoes. In: Proceedings of the second international symposium on wearable computers. Pittsburgh, PA, USA, 19–20 October 1998. IEEE, pp 132–139
84. Sun C, Shi J, Bayerl DJ, Wang X (2011) PVDF microbelts for harvesting energy from respiration. Energy Environ Sci 4:4508–4512
85. Vatansever D, Hadimani RL, Shah T, Siores E (2011) An investigation of energy harvesting from renewable source with PVDF and PZT. Smart Mater Struct 20:055019
86. Patel I, Siores E, Shah T (2010) Utilisation of smart polymers and ceramic based piezoelectric materials for scavenging waste energy. Sens Actuators A 159:213–218
87. Bauer S, Gerhard-Multhaupt R, Sessler GM (2004) Ferroelectrets: soft electroactive foams for transducers. Phys Today 57:37–43
88. Anton SR, Farinholt KM (2012) An evaluation on low-level vibration energy harvesting using piezoelectret foam. Proc SPIE 8341:83410G
89. Mellinger A (2003) Dielectric resonance spectroscopy: a versatile tool in the quest for better piezoelectric polymers. IEEE Trans Dielectr Electr Insul 10:842–861
90. Wegener M, Tuncer E, Gerhard-Multhaupt R, Bauer S (2006) Elastic properties and electromechanical coupling factor of inflated polypropylene ferroelectrets. In: 2006 IEEE Conference on Electrical Insulation and Dielectric Phenomena, Kansas City, MO, 15–18 October 2006. IEEE, Kansas, pp 752–755
91. Qi Y, Jafferis NT, Lyons K Jr, Lee CM, Ahmad H, McAlpine MC (2010) Piezoelectric ribbons printed onto rubber for flexible energy conversion. Nano Lett 10:524–525
92. Qi Y, Nguyen TD, Lisko B, Purohit PK, McAlpine CM (2011) Enhanced piezoelectricity and stretchability in energy harvesting devices fabricated from buckled PZT ribbons. Nano Lett 11:1331–1336
93. Nguyen TD, Deshmukh N, Nagarah JM, Kramer T, Purohit PK, Berry MJ, McAlpine MC (2012) Piezoelectric nanoribbons for monitoring cellular deformations. Nat Nanotechnol 7:587–593
94. Lee BY, Zhang J, Zueger C, Chung W-J, Yoo SY, Wang E, Meyer J, Ramesh R, Lee S-W (2012) Virus-based piezoelectric energy generation. Nat Nanotechnol 7:351–356
95. Glynne-Jones P, Beeby SP, White NM (2001) Towards a piezoelectric vibration-powered microgenerator. IEE Proc: sci, Meas Technol 148:68–72
96. Sterken T, Baert K, Van Hoof C, Puers R, Borghs G, Fiorini P (2004) Comparative modelling for vibration scavengers. Proc IEEE Sensors 3:1249–1252
97. Ottman GK, Hofmann HF, Lesieutre GA (2003) Optimized piezoelectric energy harvesting circuit using step-down converter in discontinuous conduction mode. IEEE Trans Power Electron 18:696–703
98. Lesieutre GA, Ottman GK, Hofmann HF (2004) Damping as a result of piezoelectric energy harvesting. J Sound Vib 269:991–1001
99. Lefeuvre E, Badel A, Richard C, Guyomar D (2005) Piezoelectric energy harvesting device optimization by synchronous electric charge extraction. J Intell Mater Syst Struct 16:865–876
100. Shu YC, Lien IC (2006) Efficiency of energy conversion for a piezoelectric power harvesting system. J Micromech Microeng 16:2429–2438
101. Lefeuvre E, Badel A, Richard C, Guyomar D (2007) Energy harvesting using piezoelectric materials: case of random vibrations. J Electroceram 19:349–355
102. Jayachandran KP, Guedes JM, Rodrigues HC (2010) Optimal configuration of microstructure in in ferroelectric materials by stochastic optimization. J Appl Phys 108:024101
103. Friswell MI, Sondipon A (2010) Sensor shape design for piezoelectric cantilever beams to harvest vibration energy. J Appl Phys 108:014901

104. Dietl JM, Garcia E (2010) Beam shape optimization for power harvesting. J Intell Mater Syst Struct 21:633–646
105. Benasciutti D, Moro L, Zelenika S, Brusa E (2010) Vibration energy scavenging via piezoelectric bimorphs of optimized shapes. Microsyst Technol 16:657–668
106. Paquin S, St-Amant Y (2010) Improving the performance of a piezoelectric energy harvester using a variable thickness beam. Smart Mater Struct 19:105020
107. Goldschmidtboeing F, Woias P (2008) Characterization of different beam shapes for piezoelectric energy harvesting. J Micromech Microeng 18:104013
108. Wang L (2011) Design, fabrication and experimental analysis of piezoelectric energy harvesters with non-traditional geometries. Thesis. Rutgers, The State University of New Jersey. New Brunswick, New Jersey
109. Arrieta AF, Hagedorn P, Erturk A, Inman DJ (2010) A piezoelectric bistable plate for nonlinear broadband energy harvesting. Appl Phys Lett 97:104102
110. Park J, Lee S, Kwak BM (2012) Design optimization of piezoelectric energy harvester subject to tip excitation. J Mech Sci Technol 26:137–143
111. Richards CD, Anderson MJ, Bahr DF, Richards RF (2004) Efficiency of energy conversion for devices containing a piezoelectric component. J Micromech Microeng 14:17–21
112. Gonsalez CG, Shiki SB, Brennan MJ, Da Silva D, Juni VL (2010) Piezoelectric energy harvesting system optimization. In: 2nd international conference on engineering optimization. Lisbon, Portugal, September 6–9, 2010. Lisbon, pp 1–8
113. Benkhelifa E, Moniri M, Tiwari A, De Rueda AG (2011) Evolutionary multi-objective design optimisation of energy harvesting MEMS: the case of a piezoelectric. In: 2011 IEEE congress of evolutionary computation, CEC 2011. New Orleans, LA, USA, 5 June 2011 through 8 June 2011. IEEE, New Orleans, pp 1856–1863
114. Bourisli RI, Al-Ajmi MA (2010) Optimization of smart beams for maximum modal electromechanical coupling using genetic algorithms. J Intell Mater Syst Struct 21:907–914
115. Hadas Z, Kurfurst J, Ondrusek C, Singule V (2012) Artificial intelligence based optimization for vibration energy harvesting applications. Microsyst Technol 18:1003–1014
116. Gurav SP, Kasyap A, Sheplak M, Cattafesta L, Haftka RT, Goosen JFL, Van Keulen F (2004) Uncertainty-based design optimization of a micro piezoelectric composite energy reclamation device. 10th AIAA/ISSMO multidisciplinary analysis and optimization conference. Albany, NY USA, 30 August 2004 through 1 September 2004.vol 6, pp 3559–3570
117. Bendsøe MP, Lund E, Olhoff N, Sigmund O (2005) Topology optimization – broadening the areas of application. Control Cybern 34:7–35
118. Bendsøe MP, Sigmund O (2003) Topology optimization−theory, methods and applications. Springer, Berlin
119. Allaire G, Jouve F, Toader AM (2004) Structural optimization using sensitivity analysis and a level-set method. J Comput Phys 194:363–393
120. Dunning PD, Kim HA (2013) A new method for creating holes in level-set function based topology optimisation. Int J Numer Meth Eng 93:118–134
121. Nakasone PH, Silva ECN (2009) Design of piezoelectric energy harvesting devices and laminate structures by applying topology optimization. Proc SPIE 7286:728603
122. Kim JE, Kim DS, Kim YY (2010) Multi-physics interpolation for the topology optimization of piezoelectric systems. Comput Methods Appl Mech Eng 199:3153–3168
123. Noh JY, Yoon GH (2012) Topology optimization of piezoelectric energy harvesting devices considering static and harmonic dynamic loads. Adv Eng Softw 53:45–60
124. Chen S, Gonella S, Chen W, Liu WK (2010) A level set approach for optimal design of smart energy harvesters. Comput Methods Appl Mech Eng 199:2532–2543
125. Merriman B, Bence JK, Osher S (1994) Motion of multiple junctions: a level set approach. J Comput Phys 112:334–363
126. Sun KH, Kim YY (2010) Layout design optimization for magneto-electro-elastic laminate composites for maximized energy conversion under mechanical loading. Smart Mater Struct 19:055008

127. Rupp CJ, Evgrafov A, Maute K, Dunn ML (2009) Design of piezoelectric energy harvesting systems: a topology optimization approach based on multilayer plates and shells. J. Int. Mater. Sys. Struct. 20:1923–1939
128. Wein F, Weller E, Albach T, Sutor A, Reinhard L (2009) Topology optimisation of a piezoelectric energy harvester. In: Sensor and test conference, proceedings sensor 2009, Nürnberg 26–28 May 2009. V.II, pp 201–206
129. Zheng B, Chang CJ, Gea HC (2009) Topology optimization of energy harvesting devices using piezoelectric materials. Struct Multi Optim 38:17–23
130. Vatanabe SL, Paulino GH, Silva ECN (2012) Influence of pattern gradation on the design of piezocomposite energy harvesting devices using topology optimization. Composites Pt B 43:2646–2654
131. Lin ZQ, Gea HC, Liu ST (2011) Design of piezoelectric energy harvesting devices subjected to broadband random vibrations by applying topology optimization. Acta Mech Sin 27:730–737
132. Wein F, Kaltenbacher M, Stingl M (2013) Topology optimization of a cantilevered piezoelectric energy harvester using stress norm constraints. Struct Multi Optim 48:173–185

Chapter 5
Conclusions

> Piezoelectrics with electromechanical coupling, shape-memory materials that can "remember" their original shape, electrorheological fluids with adjustable viscosities, and chemical sensors which act as synthetic equivalents to the human nose are examples of smart electroceramics. "Very smart" materials, in addition to sensing and actuating, have the ability to "learn" by altering their property coefficients in response to the environment. Integration of these different technologies into compact, multifunction packages is the ultimate goal of research in the area of smart material.
>
> R. E. Newnham

5.1 From Materials to Applications

The present monograph has been devoted to the performance of modern piezoelectric materials that can be applied as active elements of energy-harvesting devices or systems. In the last decade piezoelectric materials (mainly poled FCs and piezo-active composites based on either FCs or relaxor-ferroelectric SCs) have been the focus of many studies on energy-harvesting characteristics. The growing demand for effective energy-harvesting materials for the conversion of mechanical vibrations into electrical energy has motivated the production of this monograph to present the current state of knowledge in the field of piezoelectric materials, to discuss their performance and to develop concepts from previous monographs [1, 2] by the authors.

The high piezoelectric performance, strong electromechanical coupling, large figures of merit and considerable anisotropy enable us to regard the piezoelectric materials as an important group of energy harvesting materials that are suitable for effective conversion of mechanical energy into electrical energy. Based on these factors, one can select the appropriate piezoelectric materials for specific transducer, hydroacoustic, and related applications [3–5]. Piezoelectrics can also be selected for electromechanical and energy-harvesting systems [6–8] by taking into account

C.R. Bowen et al., *Modern Piezoelectric Energy-Harvesting Materials*,
Springer Series in Materials Science 238, DOI 10.1007/978-3-319-29143-7_5

modern energy-harvesting methods [9] and advanced technologies for processing novel materials [10–12].

As follows from numerous publications from the past decade, the overwhelming majority of piezoelectric energy harvesting materials are ferroelectrics or composites based on them. A unique charactristic of ferroelectric materials is the ability to combine ferroelectric, pyroelectric and electromechanical properties along with the opportunity to further improve these properties using the composite approach. These advantages make ferroelectric materials and their composites a promising and high-performance class of material for piezoelectric energy harvesting.

Our study enables us to conclude that the electromechanical properties of the piezoelectric materials and the optimistation of their properties (Fig. 5.1) open up new opportunities to use these materials for energy harvesting applications. Of particular interest are piezo-active composites with specific microgeometries and predictable effective properties [1, 2] that can be varied in wide ranges. These properties and related parameters have been determined for specific composite architectures, connectivity and components. In a number of cases, composites with 1–3, 2–2 or 0–3 connectivity patterns have the effective electromechanical properties and related parameters which can be improved by using relaxor-ferroelectric

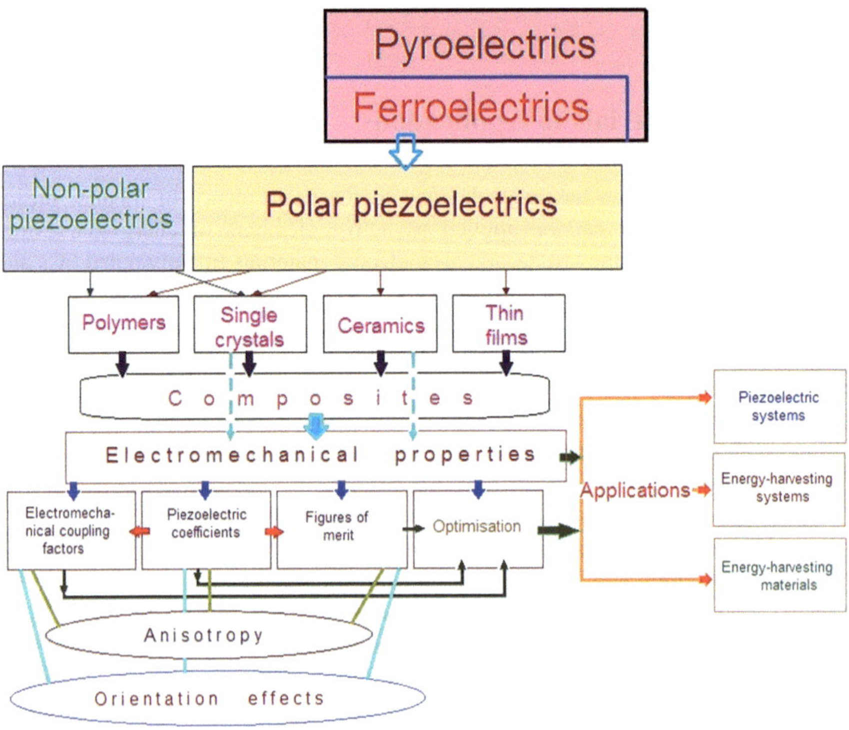

Fig. 5.1 Outline of directions of the study on piezo-active materials for energy harvesting applications

SCs with engineered-domain structures and very high piezoelectric activity, by an incorporation of porous or auxetic polymer matrices, etc. An additional stimulus to improve the electromechanical properties and parameters for piezoelectric energy harvesting is concerned with the third component and orientation effects [2] in modern piezo-active composites.

Undoubtedly, the diagram shown in Fig. 5.1 will be enlarged in the future due to new and pioneering work in the field of piezoelectric energy harvesting. It seems to be realistic that research in piezoelectric energy harvesting and related subjects (materials, properties and their optimisation) will continue to be actively carried out in the nearest future, and results from studies in physics, chemistry, mechanics, materials science, and engineering sciences will be taken into account and improved upon. It is believed that the FCs and composites based on ferroelectrics will attract interest for numerous piezotechnical and energy harvesting applications in the 21st century. In this context we would like to finish our monograph by the following words by Karl Ernst von Baer: "Die Wissenschaft ist ewig in ihrem Quell, unermesslich in ihrem Umfang, endlos in ihrer Aufgabe, unerreichbar in ihrem Ziel[1]".

References

1. Topolov VYu, Bowen CR (2009) Electromechanical properties in composites based on ferroelectrics. Springer, London
2. Topolov VYu, Bisegna P, Bowen CR (2014) Piezo-active composites. Orientation effects and anisotropy factors. Springer, Berlin
3. Safari A, Akdogan EK (eds) (2008) Piezoelectric and acoustic materials for transducer applications. Springer, New York
4. Sverdlin GM (1990) Applied hydroacoustics. Soodostroyeniye, Leningrad (in Russian)
5. Sherman CH, Butler JL (2007) Transducers and arrays for underwater sound. Springer, New York
6. Preumont A (2006) Mechatronics. Dynamics of electromechanical and piezoelectric systems. Springer, Dordrecht
7. Lenk A, Ballas RG, Werthschützky R, Pfeifer G (2011) Electromechanical systems in microtechnology and mechatronics. electrical, mechanical and acoustic networks, their interactions and applications. Springer, Berlin
8. Kaźmierski TJ, Beeby S (eds) (2011) Energy harvesting systems. Principles, modeling and applications. Springer, New York Dordrecht Heidelberg London
9. Elvin N, Erturk A (eds) (2013) Advances in energy harvesting methods. Springer, New York
10. Wang ZL, Wu W (2012) Nanotechnology-enabled energy harvesting for self-powered micro-/nanosystems. Angew Chem 51:11700–11721
11. Crossley S, Kar-Narayan S (2015) Energy harvesting performance of piezoelectric ceramic and polymer nanowires. Nanotechno 26:344001
12. Liu L, Lu K, Wang Y, Liao F, Peng M, Shao M (2015) Flexible piezoelectric nanogenerators based on silicon nanowire/α-quartz composites for mechanical energy harvesting. Mater Lett 160:222–226

[1]Science is eternal in its source, immeasurable in its scope, endless in its task, unattainable in its aim.

Appendix A
List of Abbreviations

CMOS	Complementary metal-oxide-semiconductor
DC	Direct electric current
ECF	Electromechanical coupling factor
FC	Ferroelectric ceramic
GA	Genetic algorithm
KNN-T	$(K_{0.562}Na_{0.438})(Nb_{0.768}Ta_{0.232})O_3$
KNN-TL	$Li_x(K_{0.501}Na_{0.499})_{1-x}](Nb_{0.660}Ta_{0.340})O_3$
MEMS	Micro-electro-mechanical system
MOGA	Genetic algorithm for multiple objectives
PCR	Piezoelectric ceramic from Rostov-on-Don (Russia)
PE-*n*	Auxetic polyethylene, nine modifications (n = 1, 2, …, 9)
PIN–*x*–*y*	Relaxor-ferroelectric solid solutions of $x\text{Pb}(\text{In}_{1/2}\text{Nb}_{1/2})\text{O}_3 - y\text{Pb}\cdot\cdot(\text{Mg}_{1/3}\text{Nb}_{2/3})\text{O}_3 - (1 - x - y)\text{PbTiO}_3$
PMN–*x*PT	Relaxor-ferroelectric solid solutions of $(1 - x)\text{Pb}(\text{Mg}_{1/3}\text{Nb}_{2/3})\text{O}_3 - x\text{PbTiO}_3$
PVDF	Polyvinylidene fluoride
PZN–*y*PT	Relaxor-ferroelectric solid solutions of $(1 - x)\text{Pb}(\text{Zn}_{1/3}\text{Nb}_{2/3})\text{O}_3 - x\text{PbTiO}_3$
PZT	Piezoelectric ceramic of the $\text{Pb}(\text{Zr, Ti})\text{O}_3$ type
SC	Single crystal
SIMP	Solid isotropic material with penalisation
SOMA	Self-organisation migrating algorithm

C.R. Bowen et al., *Modern Piezoelectric Energy-Harvesting Materials*,
Springer Series in Materials Science 238, DOI 10.1007/978-3-319-29143-7

Appendix B
Electromechanical Constants of Components

To analyse the energy-harvesting characteristics of piezoelectric materials and to predict the effective properties and related parameters of a piezo-active composite, we use full sets of electromechanical constants of components. A systematisation of data on the components is given in Table B.1.

Table B.1 Electromechanical constants of components at room temperature

Component	Composition	Set of constants	Table number
SC	PMN–0.33PT[a]	s^E_{ab}, d_{ij} and ε^σ_{pp}	1.1
	PMN–0.30PT[a]	s^E_{ab}, d_{ij} and ε^σ_{pp}	1.1
	PMN–0.28PT[a]	s^E_{ab}, d_{ij} and ε^σ_{pp}	1.1
	PMN–0.28PT[a,b,c]	s^E_{ab}, d_{ij} and ε^σ_{pp}	3.4
	PMN–0.42PT[b]	s^E_{ab}, d_{ij} and ε^σ_{pp}	2.15
	PZN–0.08PT[a]	s^E_{ab}, d_{ij} and ε^σ_{pp}	1.1
	PZN–0.07PT[a]	s^E_{ab}, d_{ij} and ε^σ_{pp}	1.1
	PZN–0.045PT[a]	s^E_{ab}, d_{ij} and ε^σ_{pp}	1.1
	PIN–0.24–0.49[b]	s^E_{ab}, d_{ij} and ε^σ_{pp}	3.7
	KNN-T[a]	s^E_{ab}, d_{ij} and ε^σ_{pp}	2.12
	KNN-TL[a]	s^E_{ab}, d_{ij} and ε^σ_{pp}	2.12
	$BaTiO_3$[b]	s^E_{ab}, d_{ij} and ε^σ_{pp}	2.2
	$BaTiO_3$[d]	s^E_{ab}, d_{ij} and ε^σ_{pp}	2.3
	$PbTiO_3$[b]	s^E_{ab}, d_{ij} and ε^σ_{pp}	2.2
	α-ZnS	s^E_{ab}, d_{ij} and ε^σ_{pp}	2.1
	CdS	s^E_{ab}, d_{ij} and ε^σ_{pp}	2.1
	CdSe	s^E_{ab}, d_{ij} and ε^σ_{pp}	2.1
	ZnO	s^E_{ab}, d_{ij} and ε^σ_{pp}	2.1

(continued)

C.R. Bowen et al., *Modern Piezoelectric Energy-Harvesting Materials*,
Springer Series in Materials Science 238, DOI 10.1007/978-3-319-29143-7

Table B.1 (continued)

Component	Composition	Set of constants	Table number
FC	Compositions with perovskite-type structure ($BaTiO_3$, PZT, PCR, ZTS, modified $PbTiO_3$, PMN-0.35PT, etc.)	s_{ab}^E, d_{ij} and ε_{pp}^σ	1.2
	Modified $PbTiO_3$	s_{ab}^E, d_{ij} and ε_{pp}^σ	1.2
	$(Pb_{0.80}Ca_{0.20})TiO_3$[f]	s_{ab}^E, d_{ij} and ε_{pp}^σ	3.10
	$(Pb_{0.75}Ca_{0.25})TiO_3$[f]	s_{ab}^E, d_{ij} and ε_{pp}^σ	3.10
Polymer	Araldite[e]	s_{ab} and ε_{pp}	2.7
	Polyurethane[e]	s_{ab} and ε_{pp}	2.7
	Elastomer[e]	s_{ab} and ε_{pp}	2.7
	Auxetic polyethylene	s_{ab} and ε_{pp}	2.7
	PVDF	s_{ab}^E, d_{ij} and ε_{pp}^σ	3.5

[a][001]-poled domain-engineered SC
[b]Single-domain SC
[c][011]-poled domain-engineered SC
[d]Polydomain SC, calculated values
[e]Piezo-passive material
[f]Calculated values

Appendix C
Performance of Poled Ferroelectric Ceramics

The characteristics of poled FCs are often associated with their electromechanical constants, and full sets of such constants [1–6] are of interest to determine the potential applications of FCs [7–10]. In addition to the full sets of electromechanical constants, we also use combinations of them to describe the electromechanical coupling, piezoelectric anisotropy, figures of merit, electromechanical transformations, etc. The combinations of electromechanical constants are also of value to characterise piezo-active composites based on either FCs or relaxor-ferroelectric SCs. For instance, (2.6)–(2.8), (3.14), (3.15), (3.17), and (3.18), contain effective parameters that are related to the effective electromechanical properties of the piezo-active composites. Table C.1 contains the parameters that strongly depend on the piezoelectric properties of FCs and, therefore, should be taken into account when selecting FC materials for piezoelectric energy-harvesting, transducer and hydroacoustic applications.

Table C.1 Factors of the anisotropy of piezoelectric coefficients ζ_{d3j}, ratios of electromechanical constants e_{33}/c_{33}^E (in 10^{-10} C/N), d_{33}/s_{33}^E and d_{31}/s_{11}^E (in C/m^2), ratios of coefficients of electromechanical transformation r_N, and ratios of specific acoustic powers r_{pow} at the longitudinal and transversal piezoelectric responses of perovskite-type FCs at room temperature

FC	ζ_{d3j}	e_{33}/c_{33}^E	d_{33}/s_{33}^E	d_{31}/s_{11}^E	r_N	r_{pow}
$BaTiO_3$ (I)	−2.42	1.15	21.4	−9.24	−2.32	6.12
$BaTiO_3$ (II)	−2.44	1.20	20.0	−8.57	−2.34	6.22
$(Ba_{0.917}Ca_{0.083})TiO_3$	−2.53	0.900	15.4	−6.42	−2.40	6.75
TBK-3	−2.48	0.817	14.4	−5.90	−2.41	6.26
TBKS	−3.27	0.543	8.06	−2.52	−3.19	11.0
NBS-1	−2.68	1.48	16.5	−6.85	−2.41	7.99
ZTS-19	−2.44	1.60	18.1	−8.34	−2.17	6.70
ZTS-24	−2.28	0.806	15.2	−7.96	−1.91	6.21
ZTSNV-1	−2.09	1.68	21.3	−11.0	−1.93	4.74
PZT-4	−2.35	1.31	18.7	−10.1	−1.85	7.00

(continued)

C.R. Bowen et al., *Modern Piezoelectric Energy-Harvesting Materials*,
Springer Series in Materials Science 238, DOI 10.1007/978-3-319-29143-7

Table C.1 (continued)

FC	ζ_{d3j}	e_{33}/c_{33}^{E}	d_{33}/s_{33}^{E}	d_{31}/s_{11}^{E}	r_N	r_{pow}
PZT-5	−2.19	1.42	19.9	−10.4	−1.91	5.50
PZT-5H	−2.46	1.99	27.2	−11.9	−2.29	6.50
PZT-7A	−2.51	0.769	10.9	−5.63	−1.93	8.18
Navy Type VI	−2.16	1.99	28.6	−16.5	−1.73	5.81
PZ 27	−2.41	1.38	18.6	−10.3	−1.81	7.73
PZ 34	−15.1	0.540	4.55	−0.508	−8.96	3.84
PCR-1, hp	−2.32	0.835	13.8	−7.60	−1.82	6.85
PCR-7, hp	−2.18	2.34	35.3	−16.3	−2.17	4.78
PCR-7M, hp	−2.17	2.49	38.8	−20.0	−1.94	5.27
PCR-8, hp	−2.23	1.17	18.6	−10.4	−1.79	6.21
PCR-8, ct	−2.24	1.54	19.3	−9.40	−2.05	5.47
PCR-13, ct	−2.15	1.04	12.4	−6.25	−1.98	5.02
PCR-21, hp	−2.29	1.37	19.8	−9.24	−2.14	5.60
PCR-63, hp	−2.33	1.05	14.3	−6.12	−2.33	5.43
PCR-73, hp	−2.26	2.03	36.6	−21.2	−1.72	6.71
$Pb(Zr_{0.54}Ti_{0.46})O_3$	−2.49	0.796	10.3	−5.19	−1.95	7.91
$Pb(Zr_{0.52}Ti_{0.48})O_3$	−2.39	1.12	13.0	−6.78	−1.93	7.08
$(Pb_{0.94}Sr_{0.06})\cdot(Ti_{0.47}Zr_{0.53})O_3$	−2.35	1.31	18.6	−10.0	−1.86	6.96
Modified $PbTiO_3$ (I)	−10.6	0.511	6.63	−0.666	−9.95	120
$(Pb_{0.9625}La_{0.025})\cdot(Ti_{0.99}Mn_{0.01})O_3$	−11.4	0.447	6.19	−0.576	−10.8	138
$(Pb_{0.85}Nd_{0.10})$ $(Ti_{0.99}$ $Mn_{0.01})O_3$	−10.5	0.535	7.81	−0.792	−9.88	117
$(Pb_{0.855}Nd_{0.11})$ $(Ti_{0,94}$ $Mn_{0.02}In_{0.04})O_3$	−9.05	0.531	7.94	−0.945	−8.40	88.2
PMN-0.35PT	−2.03	0.968	18.4	−10.1	−1.82	4.59

Notes

1. Parameters shown in the 2nd–7th columns were calculated using the full sets of electromechanical constants of poled FCs [1–6] at room temperature
2. FC samples of the PCR type have been manufactured using either the conventional technology (ct) or hot pressing (hp)
3. The anisotropy of the piezoelectric coefficients d_{3j} is characterised by ζ_{d3j}, and this factor is taken into account [6, 8] to provide the effective transformation of energy along the poling axis OX_3
4. The e_{33}/c_{33}^{E} ratio is of value when selecting piezoelectric components to promote large values of squared figures of merit $(Q_h^*)^2$ and $(Q_{33}^*)^2$ of the 1–3-type composites (see Sect. 3.3.1)
5. The d_{33}/s_{33}^{E} and d_{31}/s_{11}^{E} ratios are to be taken into account when selecting piezoelectric materials for transducers. Coefficients of electromechanical transformation N_{3j} linearly depend [7] on d_{3j}/s_{jj}^{E} and the geometric size of the parallelepiped (piezoelectric transducer element), where j = 1 and 3 (see Sect. 3.7)
6. According to (3.17), the ratio of the specific acoustic powers [7] related to the longitudinal and transversal piezoelectric responses of the poled FC sample is represented as $r_{pow} = (\zeta_{d3j})^2 s_{33}^{E}/s_{11}^{E}$, i.e., depends on the piezoelectric and elastic anisotropy factors (see Sect. 3.7)

References

1. Berlincourt DA, Cerran DR, Jaffe H (1964) Piezoelectric and piezomagnetic materials and their function in transducers. In: Mason W (ed.) Physical Acoustics. Principles and Methods. Vol. 1: Methods and Devices. Pt A. Academic Press, New York, London, pp 169–270
2. Khoroshun LP, Maslov BP, Leshchenko PV (1989) Prediction of effective properties of piezo-active composite materials. Naukova Dumka, Kiev (in Russian)
3. Xu Y (1991) Ferroelectric materials and their applications. North-Holland, Amsterdam
4. Dantsiger AYa, Razumovskaya ON, Reznitchenko LA, Grineva LD, Devlikanova RU, Dudkina SI, Gavrilyatchenko SV, Dergunova NV, Klevtsov AN (1994) Highly Effective Piezoceramic Materials (Handbook). Kniga, Rostov-on-Don (in Russian)
5. Topolov VYu, Bowen CR (2009) Electromechanical properties in composites based on ferroelectrics. Springer, London
6. Topolov VYu, Bisegna P, Bowen CR (2014) Piezo-active composites. Orientation effects and anisotropy factors. Springer, Heidelberg
7. Sverdlin GM (1990) Applied hydroacoustics. Soodostroyeniye, Leningrad (in Russian)
8. Gorish AV, Dudkevich VP, Kupriyanov MF, Panich AE, Turik AV (1999) Piezoelectric device-making. Vol. 1: Physics of Ferroelectric Ceramics. Radiotekhnika, Moscow (in Russian)
9. Safari A, Akdogan EK (eds) (2008) Piezoelectric and acoustic materials for transducer applications. Springer, New York
10. Kim M, Kim S-H, Hong S (2013) Materials and devices for MEMS piezoelectric energy harvesting. In: Elvin N, Erturk A (eds) Advances in energy harvesting methods. Springer, New York, pp 417–435

Index

C.R. Bowen et al., *Modern Piezoelectric Energy-Harvesting Materials*,
Springer Series in Materials Science 238, DOI 10.1007/978-3-319-29143-7

MIX
Papier aus verantwortungsvollen Quellen
Paper from responsible sources
FSC® C105338

If you have any concerns about our products,
you can contact us on
ProductSafety@springernature.com

In case Publisher is established outside the EU,
the EU authorized representative is:
Springer Nature Customer Service Center GmbH
Europaplatz 3, 69115 Heidelberg, Germany

Printed by Libri Plureos GmbH
in Hamburg, Germany